Soil Fertility Manual

To understand soil fertility is to understand a key to mankind's survival on this planet.

Published by

International Plant Nutrition Institute (IPNI)
3500 Parkway Lane, Suite 550
Norcross, GA 30092-2806

Website: www.ipni.net

Soil Fertility Manual

Copyright © 2006 by the International Plant Nutrition Institute (IPNI) and the Foundation for Agronomic Research (FAR)

Prior Editions and Printings:
First Printing, October 1978
Second Printing, January 1979
Third Printing, June 1979
Fourth Printing, March 1980
Fifth Printing, January 1982
Revised Printing, September 1982
Seventh Printing, February 1985
Eighth Printing, February 1985
Ninth Printing, November 1985
Revised Printing, September 1987
Eleventh Printing, June 1990
Twelfth Printing, June 1991
Revised Printing, August 1992
Fourteenth Printing, February 1993
Fifteenth Printing, August 1993
Sixteenth Printing, August 1994
Revised Printing, August 1995
Eighteenth Printing, December 1996
Nineteenth Printing, April 1999
Revised Printing, October 2003
Revised Printing, December 2006

Order from:

International Plant Nutrition Institute
3500 Parkway Lane, Suite 550
Norcross, Georgia 30092-2806 USA
Phone: 770-825-8082
Fax: 770-448-0439
E-mail: circulation@ipni.net
Website: www.ipni.net

ISBN # 0-9629598-5-5

Preface

The *Soil Fertility Manual* was first published in 1978. Since then, agriculture has seen many changes, such as adoption of conservation tillage, biotechnology, precision farming, and others. But the basic principles of soil fertility and agronomy have not changed. The underlying concepts of efficient nutrient management remain the same.

The purpose of this manual is to help fertilizer dealers, crop advisers, Extension workers, consultants, teachers, and agronomists give farmers sound agronomic advice. The manual provides the reader with a working knowledge of essential soil-plant relationships, principles of nutrient management and the use of aglime, and a discussion of the unique balance of agriculture and the environment. It provides answers to common questions, and studied carefully, will motivate the reader to further learning and understanding about the fundamentals of crop production.

The manual is useful as a self-study guide and can provide the basis for short courses or soil fertility workshops. It acts as a supplement to university textbooks and is an excellent resource for those preparing for the International Certified Crop Adviser (CCA) exams.

When the manual was first printed, no one could predict that it would be so well received and be such a useful tool. All of us who work in agronomy owe a debt of gratitude to Dr. Bobby C. Darst, who drafted the first version in 1978 and was involved in later updates and revisions.

We hope you and other users of this publication will find it useful and beneficial in improving understanding and appreciation of the importance of modern soil fertility.

Dr. Terry L. Roberts
President, IPNI

Acknowledgment

The International Plant Nutrition Institute, a not-for-profit agronomic research and education organization, acknowledges the support of its member companies and other cooperators in making this manual possible.

IPNI Founding Members

- Agrium
- Arab Potash Company
- Belarusian Potash Company
- Bunge Fertilizantes S.A.
- CF Industries Holdings, Inc.
- Groupe OCP
- Intrepid Mining, LLC
- K+S KALI GmbH
- Mosaic
- PotashCorp
- Saskferco
- Simplot
- Sinofert Holdings Limited
- SQM
- Terra Industries Inc.
- Uralkali

Table of Contents

Concepts of Soil Fertility and Productivity

Understanding the principles of soil fertility is vital to efficient crop production and environmental protection. There are 17 chemical elements known to be essential for plant growth. Soil texture and structure influence the amount of air and water growing plants can secure. Soil colloids are very small, negatively charged particles of minerals and organic matter. An element with an electrical charge is called an ion. Cations have positive charges; anions have negative charges. Cation exchange capacity (CEC) is an expression of the amount of a soil's negative charge and the amount of cations it can hold. Soil organic matter has many important benefits. Other key factors influencing soil productivity include soil depth, surface slope, soil organisms, and nutrient balance.

Introduction

Soils are the medium in which crops grow to feed and clothe the world. To understand soil fertility is to understand a basic need of crop production.

- How can a farmer grow profitable crops, protect the environment, and maintain long-term productivity without fertile soils?

- How can agricultural advisers help the farmer and supply information without understanding basic soil fertility?

Although soil fertility is vital to a productive soil, a fertile soil is not necessarily productive. Poor drainage, insects, drought, and other factors can limit production, even when fertility is adequate. To fully understand soil fertility, we must know other factors which support...or limit...productivity.

To understand soil productivity, one must recognize existing soil-plant relationships. Certain external factors control plant growth: air, heat (temperature), light, mechanical support, nutrients, and water. The plant depends on the soil (at least partly) for all these factors, except light. Each directly affects plant growth. Each is linked to the others. Since water and air occupy the pore spaces in the soil, factors that affect water relations necessarily influence soil air. In turn, moisture changes affect soil temperature. Nutrient availability is influenced by soil and water balance as well as by soil temperatures. Root growth is also influenced by soil temperature as well as soil water and air.

Soil fertility in modern-day agriculture is a part of a dynamic system. Nutrients are constantly being removed from the soil in the form of plant and animal products. Unfortunately, others can be lost by leaching or erosion. Still others, like phosphorus (P) and potassium (K), can be tied up by certain soil clays. Organic matter and soil organisms immobilize, then release, nutrients. If production agriculture were a closed system, nutrient balance might be relatively stable. It's not, however, and that is why understanding the principles of soil fertility is essential to efficient crop production and environmental protection.

This chapter features soil traits that influence plant growth. Essential plant nutrients are also listed and categorized.

The following chapters contain production information on all the essential plant nutrients, including amounts good crop yields remove, their roles in plant growth, their deficiency symptoms in plants, their soil relationships, their fertilizer sources, and their impacts on the environment.

Essential Plant Nutrients

Seventeen chemical elements are known to be essential for plant growth. An element is considered essential if it is necessary for the plant to complete its life cycle, including vegetative and reproductive phases, and no other element can substitute for it completely. Essential nutrients are divided into two main groups: non-mineral and mineral.

Non-mineral nutrients are carbon (C), hydrogen (H), and oxygen (O). These nutrients are found in the atmosphere and water. They are used in photosynthesis in this manner:

$$6CO_2 + 12H_2O \xrightarrow{\text{Light}} 6O_2 + 6(CH_2O) + 6H_2O$$

| Carbon dioxide | Water | | Oxygen | Carbohydrates | Water |

Products of photosynthesis account for most of the increase in plant growth. Insufficient carbon dioxide, water, or light reduce growth. However, the amount of water used in photosynthesis is so small that plants will show moisture stress before water is low enough to affect photosynthesis rate. See **Production Concept 1-1.**

The 14 mineral nutrients...those coming from the soil...are divided into three groups: primary, secondary, and micronutrients:

Primary Nutrients
Nitrogen (N)
Phosphorus (P)
Potassium (K)

Secondary Nutrients
Calcium (Ca)
Magnesium (Mg)
Sulfur (S)

Micronutrients
Boron (B)
Chloride (Cl)
Copper (Cu)
Iron (Fe)
Manganese (Mn)
Molybdenum (Mo)
Nickel (Ni)
Zinc (Zn)

Four additional nutrients—sodium (Na), cobalt (Co), vanadium (V), and silicon (Si)—have also been established as essential micronutrients in some plants. They are almost never deficient in soils.

The primary nutrients usually become deficient in the soil

first, because plants use relatively large amounts. The secondary nutrients and micronutrients are usually deficient less often. Although smaller amounts are usually used, they are just as important as the primary nutrients. Plants must have secondary nutrients and micronutrients when and where they need them in meeting the nutritional needs of crops.

Soil Texture and Structure

Soil texture is determined by how much sand, silt, and clay are in the soil. Sand particles range in diameter from 2.0 to 0.05 millimeters (mm), silt from 0.05 to 0.002 mm, and clay particles are less than 0.002 mm in diameter. Sand, silt, and clay are often referred to as primary soil particles.

In terms of texture, the smaller the particle size, the closer to clay; the larger the particle size, the closer to sand. For example:

- A soil high in sand content is classified in texture as "sand."

- When small amounts of silt or clay are present, the soil becomes either "loamy sand" or "sandy loam."

- Soils composed mostly of clays are "clay."

- When sand, silt, and clay are present in similar proportions, the soil is called a "loam."

The 12 textural classes for soils are shown in **Figure 1-1**.

Soil structure is the arrangement of primary soil particles into secondary units or peds. Examples of structural types are

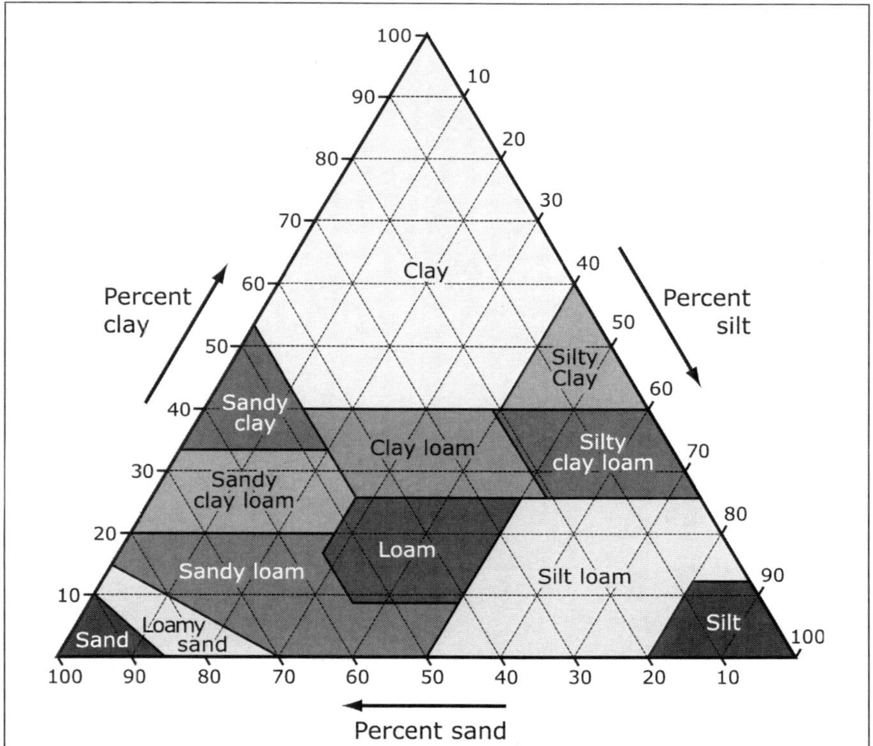

Figure 1-1. Textural chart identifying soil types by sand, silt, and clay contents.

Photosynthesis– Nature's Miracle

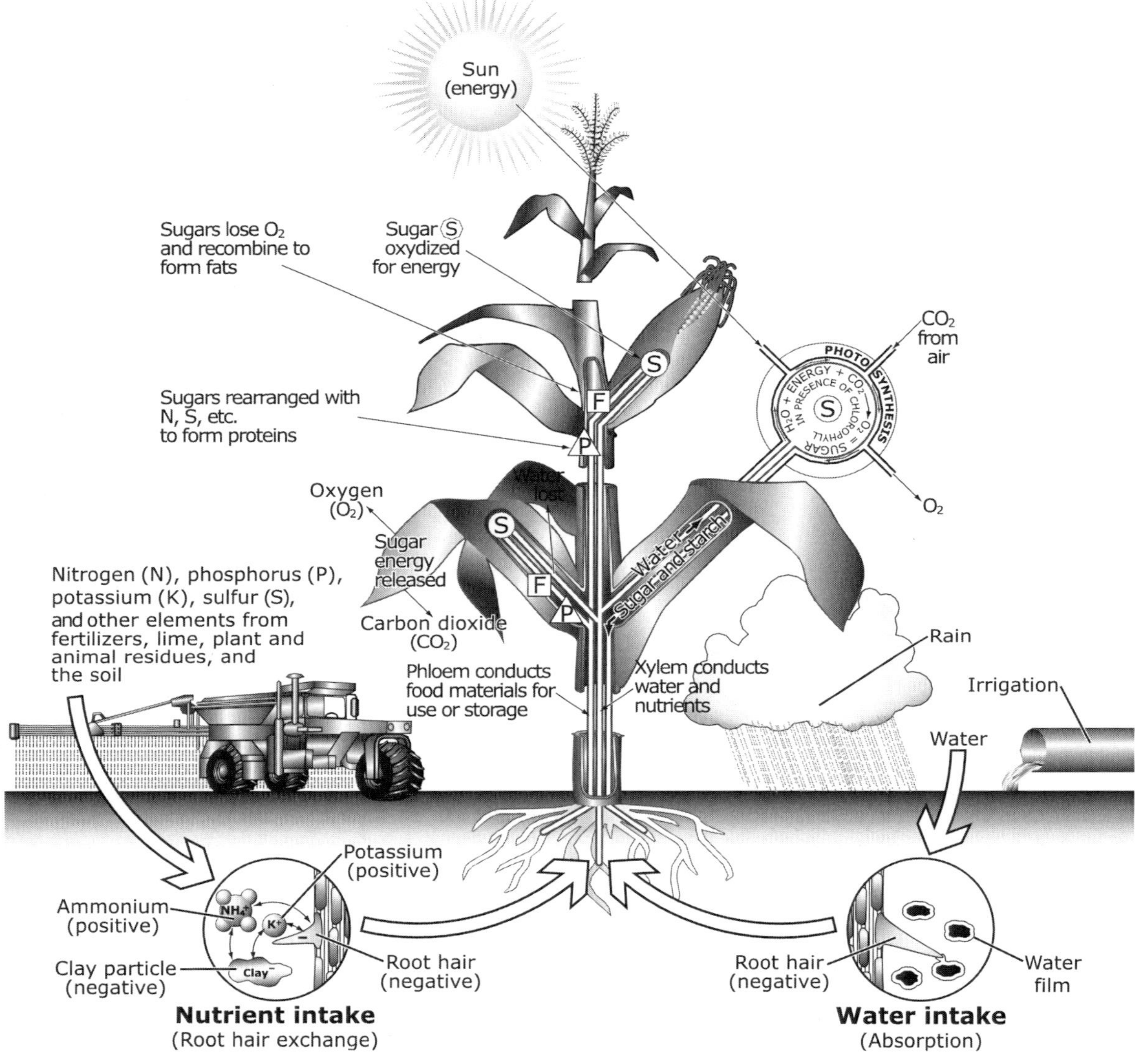

This model shows how a plant uses water and nutrients from the soil and O_2 from the air to manufacture sugars (S), fats (F), and proteins (P). The more it can manufacture, the more food or fiber that plant can yield.

High-yield agriculture provides several great helps to nature: 1) more N, P, K, lime, and other nutrients needed to assure an adequate supply for optimum yields; 2) water control through irrigation and/or drainage or soil practices that improve water use; 3) good tillage and production practices to seek the best possible growing environment; 4) more land undisturbed for wildlife habitat and recreational opportunities for people.

blocky, granular, and prismatic.

Soil texture and structure influence the amount of air and water growing plants can use. Particle size is important because:

- Smaller clay particles are more tightly fitted together than larger sand particles. This means small pores for air and water.
- Smaller particles have much higher surface areas than larger ones. Fine clay has about 10,000 times as much surface area as the same weight of medium-sized sand. As surface area increases, the amount of water adsorbed (held) increases.

Thus, sands hold little water because their large pore spaces allow water to drain freely from the soils. Clays adsorb a relatively large amount of water, and their small pore spaces retain it against gravitational forces. Although clay soils have greater waterholding capacities than sandy soils, not all the moisture is available to growing plants. Clay soils (and those high in organic matter) hold water more tightly than sandy soils. This means more unavailable water. While clay soils hold more water than sands, more of it is unavailable.

The term field capacity defines the amount of water remaining in a soil after gravitational flow has stopped. It can be expressed as a percent by weight or volume of soil. The amount of water a soil contains after plants are permanently wilted is called the permanent wilting percentage. Water is still present at this point, but is held so tightly that plants are unable to use it. Water available to growing plants is that amount contained in the soil between field capacity and permanent wilting percentage. **Figure 1-2** shows how available water varies with soil texture.

Sandy soils cannot store as much water as clay soils. But a higher percentage of that present is available in sandy soils. So there is no constant relationship between soil texture and available water, as shown in **Figure 1-2**.

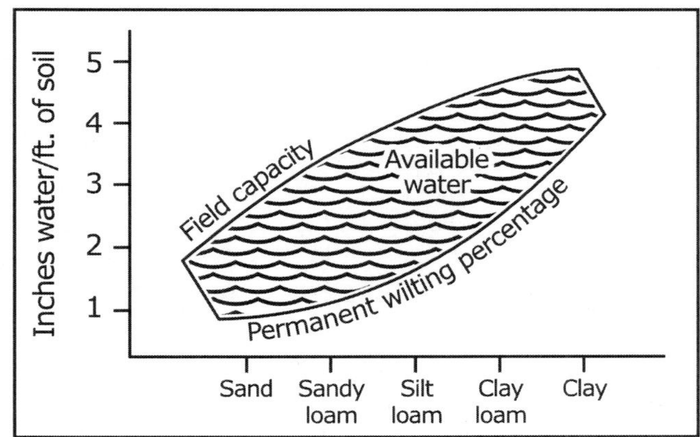

Figure 1-2. Relationship between soil texture and water availability.

Fine textured soils (clays) are easily compacted. This reduces pore space which limits air and water movement through the soil, causing much rainfall to run off. Moisture stress can become a problem, even under high rainfall. Clays are sticky when wet and form hard clods when dry. Hence, moisture content is an extremely important consideration when tillage is performed. These soils are most susceptible to compaction at or near field capacity.

Sandy soils are inherently droughty because they can hold little water. They are loose, less likely than clays to be compacted, and are easy to till. However, soils containing high proportions of very fine sand are easily compacted.

Soils high in silt are often the most difficult of all in terms of soil structure. The particles fit very closely together and may be compacted readily.

Good management helps to maintain or develop good soil structure. Size and shape of granules determine structure quality. The best structure is blocky and granular, with particles aggregated, to allow for free air and water movement.

Soil structure sharply influences root and top growth. As the soil becomes more compacted, the proportion of large pore spaces decreases, root growth bogs down, and production declines. The ideal soil for crop production can be characterized as follows:

- Medium texture (loamy) and organic matter for air and water movement.
- Sufficient clay to hold reserve soil moisture.
- Deep, permeable subsoil with adequate fertility levels.
- Environment for roots to go deep for moisture and nutrients.

Soil Colloids and Ions

As soils are formed during the weathering processes, some minerals and organic matter are broken down to extremely small particles. Chemical changes further reduce these particles until they cannot be seen with the naked eye. The very smallest are called colloids. Scientists have learned that most mineral clay colloids are plate-like in structure and crystalline in nature. Most soils contain more clay colloids than organic colloids. Colloids are primarily responsible for the chemical reactivity in soils. The kind of parent material and the degree of weathering determine the kinds of clays present in the soil. Since soil colloids are derived from these clays, their reactivity is also influenced by parent material and weathering.

Each colloid (clay and organic) has a net negative (-) charge, developed during the formation process. This means it can attract and hold positively (+) charged particles, as unlike poles of a magnet attract each other. Colloids repel other negatively charged particles, again as like poles of a magnet repel each other.

An element with an electrical charge is called an ion. Potassium, Na, H, Ca, and Mg all have positive charges. They are called cations. They are written in ionic form as shown **Table 1-1**. Note that some cations have more than one positive charge.

Table 1-1.   Common soil cations and ionic forms.

Cation	Ionic form
Potassium	K^+
Sodium	Na^+
Hydrogen	H^+
Calcium	Ca^{2+}
Magnesium	Mg^{2+}

Soil Fertility Manual

Ions with negative charges, such as nitrate (NO_3^-) and sulfate (SO_4^{2-}), are called anions. **Table 1-2** shows some common anions.

Table 1-2.   Common soil anions and ionic forms.

Anion	Ionic form
Chloride	Cl^-
Nitrate	NO_3^-
Sulfate	SO_4^{2-}
Phosphate	HPO_4^{2-}; $H_2PO_4^-$

Negatively charged colloids attract cations and hold them like a magnet holding small pieces of metal. This characteristic explains why NO_3^--N is more easily leached from the soil than ammonium (NH_4^+)-N. Nitrate has a negative charge, like soil colloids. So, NO_3^- is not held by the soil, but remains as a free ion in soil water to be leached through the soil profile in some soils and under some rainfall conditions. The concept is shown in **Figure 1-3**.

Review Questions:

11. (T or F) A soil colloid is visible to the naked eye.

12. Soil colloids have _______ charges which were developed while the soil was being formed.

13. A cation has a plus (+) or minus (-) charge (choose one).

14. An anion has a plus (+) or minus (-) charge (choose one).

15. Based on the fact that opposite charges are attracted to each other, which of the following would be attracted to a soil colloid: K^+, cation, NO_3^-, SO_4^{2-}, Ca^{2+}, anion?

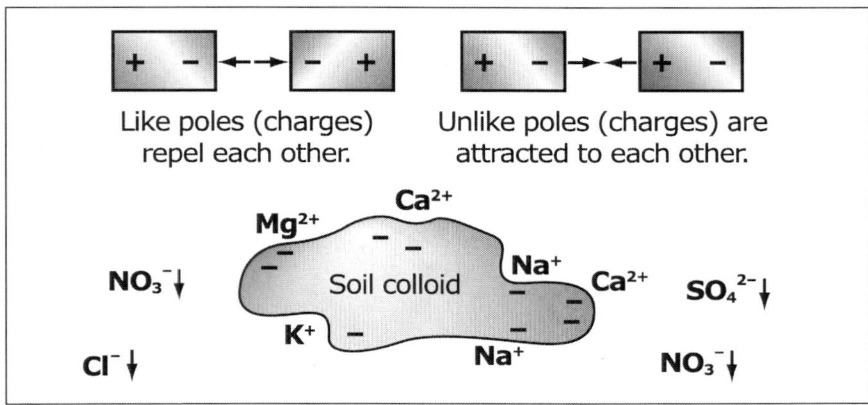

Figure 1-3.   Cations are attracted to soil clays and organic matter; anions are repelled.

Cation Exchange Capacity
(See **Production Concepts 1-2 and 1-3**)

Cations held by soils can be replaced by other cations. This means they are exchangeable. For example, Ca^{2+} can be exchanged for H^+ and/or K^+ and vice versa. The total number of exchangeable cations a soil can hold (the amount of its negative charge) is called its cation exchange capacity or CEC. The higher a soil's CEC, the more cations it can retain. Soils differ in their capacities to hold exchangeable K^+ and other cations. The CEC depends on amounts and kinds of clay and organic matter present. A high-clay soil can hold more exchangeable cations than a low-clay soil. Also, CEC increases as organic matter increases.

The CEC of a soil is expressed in terms of centimoles of charge per kilogram of soil and is written as cmol/kg. It may also be expressed as milliequivalents per 100 grams soil (meq/100 g), which is numerically equal to cmol/kg. Clay minerals usually range from about 10 to 150 cmol/kg in CEC. Organic matter ranges from 200 to 400 cmol/kg. So, the kind and amount of clay and organic matter content greatly influence the CEC of soils.

Where soils are highly weathered and organic matter levels are low, CEC values are low. Where less weathering has occurred and organic matter levels are higher, CEC values can be quite high. Clay soils with high CEC can retain large amounts of cations against potential loss by leaching. Sandy soils, with low CEC, retain smaller quantities. This makes timing and application rates important in planning a fertilizer program. For example, it may not be wise to apply K on very sandy soils in the middle of a normally wet season, where rainfall can be high and intense. Fertilizer application should be split to prevent leaching and erosion, especially in humid, high rainfall areas. Also, splitting N applications, using nitrification inhibitors, and timing applications to meet peak crop demands are important to reduce the potential for NO_3^- leaching on sands as well as finer-textured soils.

Percent base saturation...the percent of total CEC occupied by the exchangeable basic cations (Ca^{2+}, Mg^{2+}, K^+, Na^+)...has been used in the past to develop fertilizer programs. The idea is that certain nutrient ratios or balances are needed to assure proper uptake by the crop for optimum yields. Research has shown, however, that cation saturation ranges and ratios have little or no utility in a vast majority of agricultural soils. Under field conditions, ranges of nutrients can vary widely with no detrimental effects, so long as individual nutrients are present in sufficient levels in the soil to support optimum plant growth.

Anion Retention in Soil

There is no clear-cut mechanism for the retention of anions by the soil. Nitrate, for example, is completely mobile and moves freely with soil moisture. Under high rainfall, it moves downward. Under extremely dry weather, it moves upward with soil moisture, causing NO_3^- to accumulate on the surface.

Sulfate can be held, rather loosely, in some soils under certain conditions. At low pHs, positive charges can develop along broken edges of some clays. Soils containing hydrous Fe and aluminum (Al) oxides, either in topsoil or subsoil, hold some SO_4^{2-} through positive charges developed. But this small retention is of little consequence above pH 6.0. Large quantities of SO_4^{2-} can be retained through accumulation of gypsum ($CaSO_4$) in arid and semi-arid regions.

Phosphate, unlike NO_3^- and SO_4^{2-}, is held strongly by the soil and is considered immobile because of adsorption reactions (P fixation) and other interactions with soil minerals.

Sulfate salts can be held on the surface of soil colloids, and SO_4^{2-} may be loosely held by other complexes which are adsorbed. Organic matter sometimes develops a positive charge. When it does, SO_4^{2-} can be attracted to it.

Clay and Organic Matter Particles

Soil texture	Approximate percent clay
Loamy sand	5
Sandy loam	10
Silt loam	20
Silty clay loam	30
Clay loam	35
Clay	>40

To understand nutrient behavior in the soil, we must understand the role of clay and organic matter particles. All farm soils contain some clay and some organic matter. Typical clay contents of major soil classes are shown above.

The sketch below explains: (1) How cations are held by clay and organic matter to resist leaching and (2) How anions are repelled.

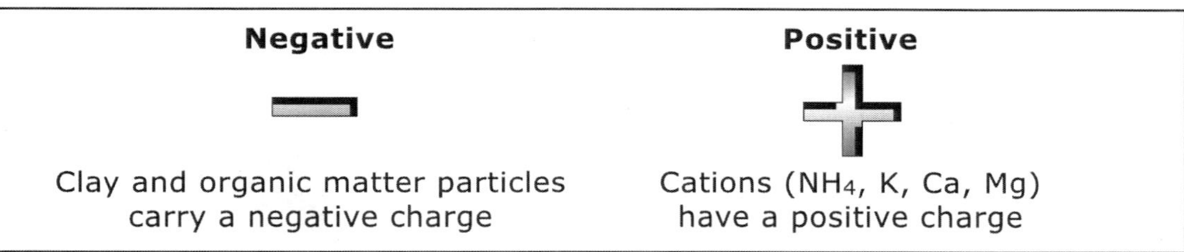

Cations are held on clay and organic matter particles by magnetic attraction.

Unlike poles attract — like poles repel. This is the same principle that holds cations to the clay and organic matter particles.

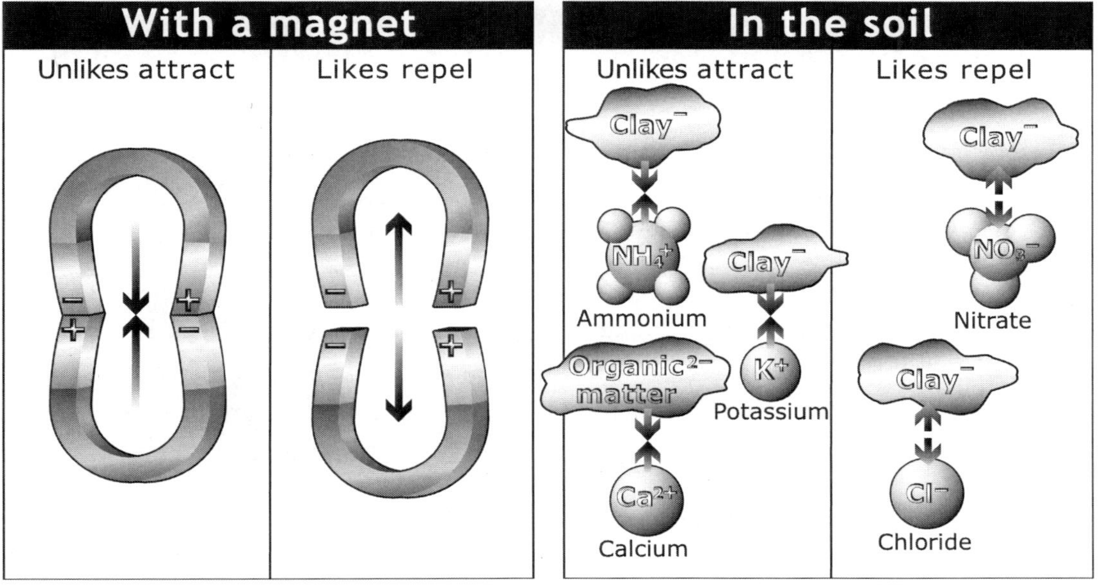

Cation Exchange Capacity:
An aid in soil management and nutrient addition

Cations are positively charged nutrient ions and molecules: calcium (Ca^{2+}), magnesium (Mg^{2+}), potassium (K^+), sodium (Na^+), hydrogen (H^+), and ammonium (NH_4^+).

Clay particles are the negatively charged constituents of soils. They attract, hold, and release positively charged nutrient particles (cations). Organic matter particles also have a negative charge to attract positively charged cations. Sand particles carry little or no charge and do not react.

Cation exchange capacity is the soil's capacity to hold and exchange cations. The strength of a cation's positive charge varies, enabling one cation to replace another on a negatively charged soil particle.

A Schematic Look at Cation Exchange

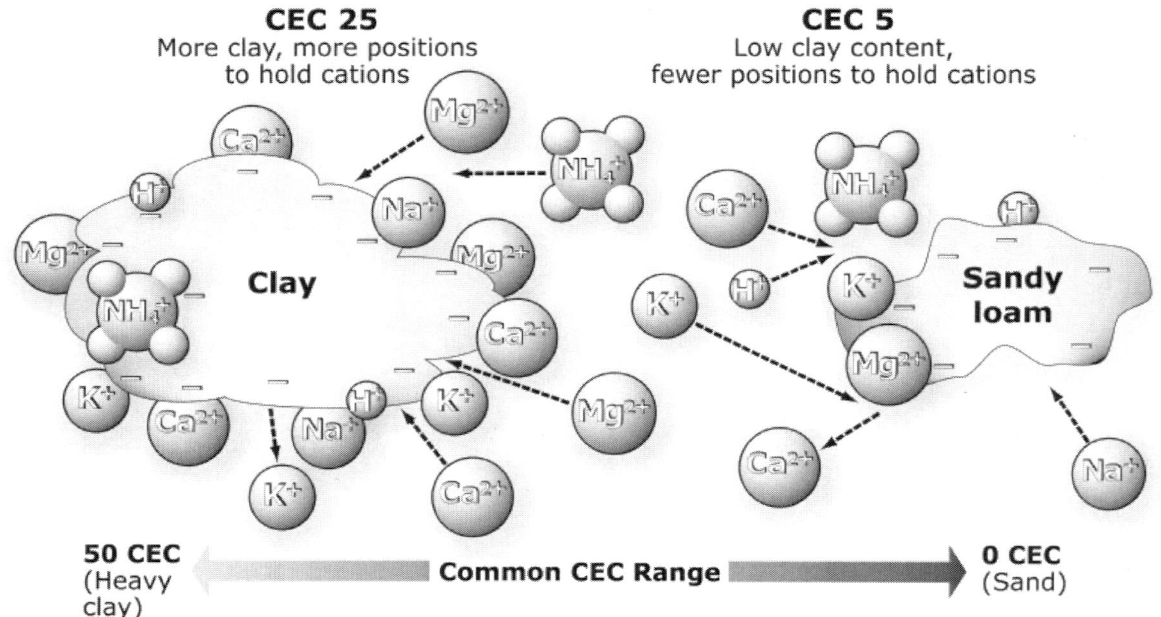

Some Practical Applications

Soils with CEC 11-50 range

- High clay content
- More aglime required to correct a given pH
- Greater capacity to hold nutrients in a given soil depth
- Physical ramifications of a soil with a high clay content
- High water-holding capacity

Soils with CEC 1-10 range

- High sand content
- Nitrogen and K leaching more likely
- Less aglime required to correct a given pH
- Physical ramifications of a soil with a high sand content
- Low water-holding capacity

Soil organic matter consists of plant, animal, and microbial residues in various stages of decay (decomposition). It benefits the soil in several ways, including the release of various products as decomposition takes place. The following are benefits of soil organic matter.

- Improves soil physical condition.

- Increases water infiltration and aeration.

- Improves soil tilth and structure by strengthening soil aggregates.

- Increases water holding capacity.

- Reduces soil erosion.

- Increases the soil's CEC.

- Supplies plant nutrients, including N, P, and S. Each 1% of organic matter can release in the range of 10 to 40 lb N/A each year.

- Reduces soil compaction and surface crusting.

- Buffers the soil against rapid changes in pH.

- Serves as an energy source for microorganisms.

Organic matter contains about 5% total N, so it serves as a storehouse for reserve N. However, the N is tied up in organic compounds and not immediately available for plant use, since decomposition usually occurs slowly. Although a soil may contain much organic matter, fertilizer N is needed to assure non-legume crops an adequate source of readily available N, especially those crops requiring high N...corn, wheat, cotton, etc. Other essential nutrients are also contained in soil organic matter. Plant and animal residues contain variable amounts of mineral elements such as P, Mg, Ca, S, and the micronutrients. As organic matter decomposes, these elements are mineralized and become available to growing plants.

Organic matter decomposition releases nutrients, but N and S can be temporarily tied up during the process. Microorganisms decomposing the organic matter require N to build protein in their bodies. If the organic matter being decomposed has a high C/N ratio...meaning low N...these organisms will use available soil and fertilizer N.

When residues of cotton and corn stalks or oat and wheat straw are incorporated into the soil, additional N should be applied if a crop is to be planted soon after that. If not, crops may suffer temporary N deficiency.

Eventually, N immobilized in the bodies of soil organisms becomes available as the organisms die and decay. With conservation tillage and the buildup of residues as crop yields increase, N management requires extra attention until a new soil equilibrium is reached. Extra care should be taken to avoid a deficiency from too little N. At the same time, rates used should not exceed crop need so that potential NO_3^- leaching can be minimized. See Chapters 3 and 11 for more detail on N management.

Some soils contain very little organic matter. In tropical areas, most soils are inherently low in organic matter because warm temperatures and high rainfall speed up decomposition. Research is showing, however, that organic matter levels can be built in these soils with management that produces higher yields and more residue per acre. In cooler areas, where decomposition takes place more slowly, native organic matter levels can be quite high. With adequate fertilization and good management practices, more crop residues are produced. In high yielding corn fields, residue amounts as much as 8 tons/A are left after the grain has been harvested. Residues help maintain or increase organic matter levels in soils. They benefit physical, chemical, and microbial soil properties. They should be added regularly to sustain crop production. The important point is to keep sufficient amounts of residues cycling through the soil.

Other Factors Affecting Soil Productivity

Soil Depth

Soil depth may be defined as that depth of soil material favorable for plant root penetration. Deep, well drained soils of desirable texture and structure are favorable to crop production. Plants need plenty of depth for roots to grow and secure nutrients and water. Roots will extend 3 to 6 ft. or more when soil permits. Alfalfa roots may reach depths of 30 ft. Rooting depth can be limited by physical and chemical barriers as well as by high water tables. Hardpans, shale beds, gravelly layers, and accumulations of soluble salts are extremely difficult to correct. But a high water table can usually be corrected with improved drainage. **Table 1-3** rates relative soil productivity by soil depth.

Table 1-3. Influence of soil depth on relative productivity.

Soil depth usable by crop roots, ft.	Relative productivity, %
1	35
2	60
3	75
4	85
5	95
6	100

Surface Slope

Topography largely determines the amount of runoff and erosion. It also dictates irrigation methods, drainage, conservation measures, and other best management practices (BMPs) needed to conserve soil and water. The steeper the land, the more management is needed, increasing labor and equipment costs. At certain slopes, soil becomes unsatisfactory for row crop production. The ease with which surface soils erode, along with percent slope, is a determining factor in a soil's potential productivity. **Table 1-4** rates relative productivity by soil slope and erodibility.

Table 1-4. Influence of soil slope on relative productivity.

Soil slope, %	Relative productivity, %[1]	
	Soil not easily eroded	Soil easily eroded
0-1	100	95
1-3	90	75
3-5	80	50
5-8	60	30

[1]Conservation tillage helps reduce the detrimental effects of slope.

Soil Organisms

Many types of organisms live in the soil. They range in size from microscopic (bacteria, nematodes, and fungi) to groups readily visible to the naked eye (earthworms and insect larvae). Some of the microscopic organisms cause many beneficial soil reactions, such as the decay of plant and animal residues. They help to speed up nutrient cycling. Other reactions are injurious, such as the development of organisms that cause plant and animal diseases. Most soil organisms depend on organic matter for food and energy, so are usually found in the upper foot of soil. Factors affecting the abundance of soil organisms include: moisture, temperature, aeration, nutrient supply, soil pH, and the type of crop being grown. Good fertilizer management, along with other BMPs, helps to maintain soil organisms at desired levels. Chapter 3 discusses activities of some types of soil organisms.

Nutrient Balance

Nutrient balance is vital to soil fertility and crop production. Although N is usually the first limiting nutrient for non-legume crops, without adequate amounts of other nutrients, N use efficiency (NUE) suffers. For example, increased N uptake and utilization with adequate K mean improved NUE and higher yields.

Figure 1-4 shows how corn yield and NUE were increased by fertilizer K (Illinois) and soil test K (Ohio), resulting in economic and environmental benefits.

Good crop growth demands proper nutrient balance. **Table 1-5** shows how P balanced with N increased yield and NUE and uptake.

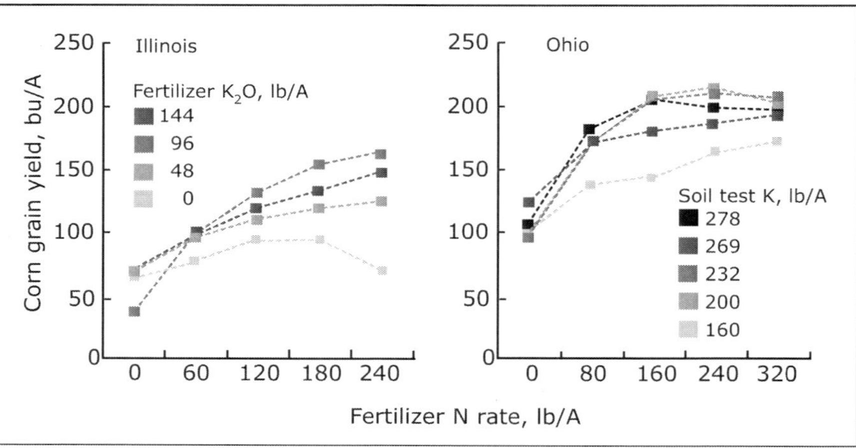

Table 1-5. Phosphorus fertilization increases ryegrass yield and NUE.

P₂O₅ rate, lb/A	Yield, lb/A	NUE, lb forage/lb N	N uptake, lb/A[1]
0	2,180	18	44
40	5,320	44	106
80	5,990	50	120
120	6,220	52	124

N rate = 120 lb/A
[1]Assumes 40 lb N/ton

Texas

Summary

Many factors control soil productivity. Fertilizer use is only one. Failure to employ sound production practices reduces the potential benefits of fertilizer and limits productivity. Understanding those factors which control productivity and applying them increases yields, improves farmer profits, and helps to protect the environment. 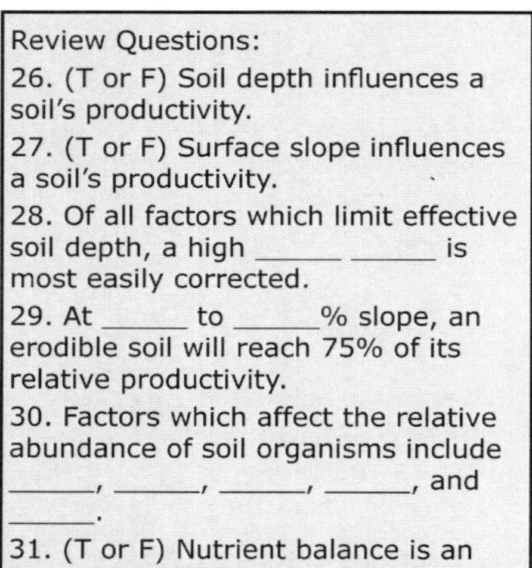

Chapter 2

Soil pH and Liming

Soil pH is an important characteristic defining relative acidity or basicity. Many factors influence soil pH, including parent material, precipitation, native vegetation, crops grown, soil depth, and nitrogen (N) fertilization. There are various methods of measuring soil pH and determining appropriate aglime requirements. When soil pH is too low (acidity is too high), various detrimental effects may depress crop growth. Desirable pH levels vary...some crops thrive in more acid conditions. The process and reactions by which aglime reduces pH are complex. Time and frequency of aglime applications are important considerations. Neutralizing value, degree of fineness, and reactivity are key considerations in quality of aglime. Placement of the liming material is also important. Several different materials can be used in liming to adjust soil acidity. Some soils in arid climates have high pHs which can affect their properties and influence productivity, thus requiring special management.

What Is Soil pH?

The term pH defines the relative acidity or basicity of a substance. The pH scale covers a range from zero to 14. A pH value of 7.0 is neutral. Values below 7.0 are acid. Those above 7.0 are basic. Most productive soils range from 5.5 to 8.0 in pH.

An acid is a substance that releases hydrogen ions (H^+). When saturated with H^+, a soil behaves as a weak acid. The more H^+ held on the exchange complex, the greater the soil's acidity. Aluminum (Al) also acts as an acidic element and activates H^+. The degrees of acidity and basicity for pHs ranging from 5 to 10 are shown in **Figure 2-1**.

Soil pH simply measures H^+ activity in moles per liter and is expressed in logarithmic terms. The practical significance of the logarithmic relationship is that each unit change in soil pH means a 10-fold change in the amount of acidity or basicity. That is, a soil with a pH of 6.0 has 10 times as much active H^+ as it would with a pH of 7.0. Therefore, aglime need increases rapidly as pH drops. **Table 2-1** shows the relative degrees of acidity and basicity, compared to a neutral pH of 7.0.

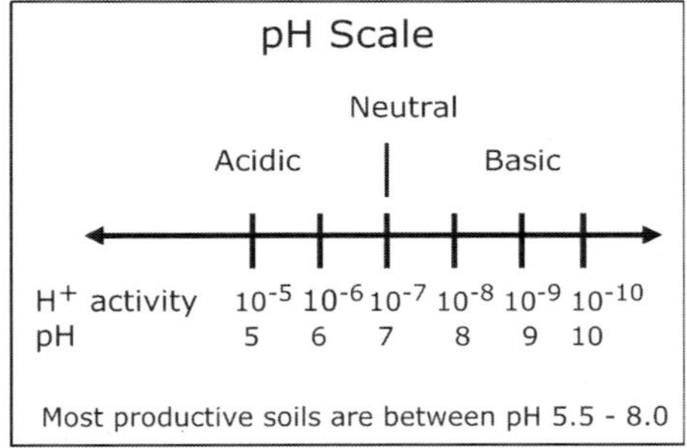

Figure 2-1. Degrees of acidity and basicity found in most agricultural soils.

Table 2-1. Comparison of magnitudes of acidity and basicity at different pH values.

Soil pH		Acidity/basicity compared to pH 7.0
9.0	Basicity	100
8.0		10
7.0	Neutrality	
6.0		10
5.0	Acidity	100
4.0		1,000

Review Questions:
1. Soil pH is a measure of __________ activity and is expressed in ________ terms.
2. (T or F) A soil with a pH of 7.5 is basic in reaction.
3. (T or F) A soil with a pH of 7.0 is acid in reaction.

Soil pH Status in North America

A recent summary conducted by the Potash & Phosphate Institute, and including about 3.4 million soil samples, showed that the median pH for the U.S. and Canada is 6.3, with 31% of the samples testing less than 6.0. A pH of 6.0 was chosen as the breaking point because a soil pH of 6.0 or greater is desirable for most cropping systems. **Figure 2-2** shows the median pH by state and province.

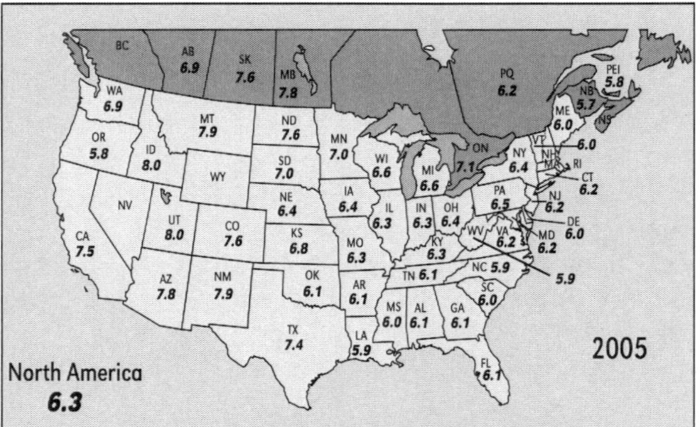

Figure 2-2. Median soil pH levels in North America; 31% of samples for 2005 were below 6.0.

Factors Affecting Soil pH

Soil pH is influenced by several factors, including decomposition of organic matter, parent material, precipitation, native vegetation, crops grown, soil depth, N fertilization, and flooding.

- **Decomposition of organic matter**–Soil organic matter is continuously being decomposed by microorganisms into organic acids, carbon dioxide (CO_2), and water, forming carbonic acid. Carbonic acid, in turn, reacts with the calcium (Ca) and magnesium (Mg) carbonates in the soil to form more soluble bicarbonates which are leached away, leaving the soil more acid.

- **Parent material**–Soils developed from parent material of basic rocks generally have higher pHs than those formed from acid rocks.

- **Precipitation**–As water from rainfall passes through the soil, basic nutrients such as Ca and Mg are leached. They are replaced by acidic elements, including Al, H, and manganese (Mn). Therefore, soils formed under high rainfall conditions are more acid than those formed under arid conditions.

- **Native vegetation**–Soils formed under forest vegetation tend to be more acid than those developed under grasslands. Conifers create more acidity than do deciduous forests.

- **Crops grown**–Soils often become more acid when crops are harvested because bases are removed. Type of crop determines the relative amounts of removal. For example, legumes generally contain higher levels of bases than grasses. Legumes also release H$^+$ ions into their rhizo-sphere, a zone of increased microbial growth and activity surrounding the roots, when actively fixing atmospheric N.

- **Soil depth**–Except in low rainfall areas, acidity generally increases with depth, so the loss of topsoil by erosion can lead to a more acid pH in the plow layer. The reason is that more subsoil is included in the plow layer as topsoil is lost. There are areas, however, where subsoil pH is higher than that of the topsoil. It is important to understand the nature of soil pH change with depth, either through practical knowledge or the use of soil testing, when planning a liming program.

- **Nitrogen fertilization**–Nitrogen...from fertilizer, organic matter, manure, and legume N fixation... produces acidity. Nitrogen fertilization speeds up the rate at which acidity develops. At lower N rates, acidification rate is slow, but is accelerated as N fertilizer rates increase. (See Chapter 3 for discussion of N fertilizer effects on soil pH.) On calcareous soils, the acidifying effect can be beneficial. When iron (Fe), Mn, or other micronutrients might be deficient, lowering pH makes them more available, except for molybdenum (Mo). Unlike most other micronutrients, the availability of Mo increases with increasing soil pH.

- **Flooding**–The overall effect of submergence is an increase of pH in acid soils and a decrease in basic soils. Regardless of their original pH values, most soils reach pHs of 6.5 to 7.2 within one month after flooding and remain at that level until dried. Consequently, liming is of little value in flooded rice production. Further, it can induce deficiencies of micronutrients such as zinc (Zn).

Determining Aglime Requirement

Although soil pH is an excellent single indicator of soil acidity, it does not determine lime requirement. Lime requirement is the amount of aglime needed to establish the desired pH range for the cropping system being used. When pH is measured, only active acidity in the soil water is determined. Potential acidity, held by soil clay and organic matter, must also be considered. Some method of relating a change in soil pH to the addition of a known amount of acid or base is necessary. Such a method is called lime requirement determination.

The lime requirement of a soil is not only related to the pH of that soil, but also to its buffer capacity or cation exchange capacity (CEC). Total amounts of clay and organic matter in a soil, as well as the kind of clay, will determine how strongly soils are buffered...how much they resist a pH change. Buffering capacity increases with the amounts of clay and organic matter. Such soils require more lime to increase pH than soils with a lower buffer capacity. Sandy soils, with small amounts of clay and organic matter, are weakly buffered, so they require less aglime to change the pH.

A common method of determining aglime need is based on the pH change of a buffered solution compared to the pH of a soil-water suspension. An acid soil will depress the pH of the buffer. The pH is depressed in proportion to the original soil pH and the soil's buffering capacity. By calibrating pH changes in the buffered solution which accompany the addition of known amounts of acid, the amount of aglime required to bring the soil to a particular pH can be determined. There are several buffer methods in use.

Regardless of the method used to determine soil pH and aglime requirement, aglime use should be based on one that is reliable. Too much aglime on coarse-textured soils can lead to excessively basic conditions and serious problems...such as Fe, Mn, and other micronutrient deficiencies. Yet, the amounts of aglime which would harm crops growing on sandy soils may not be enough to raise pH to the desired level on clay or high organic matter soils.

Why Acid Soils Should Be Limed

Soil acidity affects plant growth in many ways. See **Production Concept 2-1**. Whenever pH is low (acidity is high), one or more detrimental effects may depress crop growth.

Listed here are some consequences of low soil pH.

- Concentrations of such elements as Al, Fe, and Mn can reach toxic levels because their solubilities increase in acid soils.

- Organisms responsible for decaying organic matter and transforming N, P, and sulfur (S) may be low in number and activity.

- Calcium may be deficient (rarely) when the CEC of the soil is extremely low. Magnesium may also be deficient.

- The performance of some soil-applied herbicides can be adversely affected when soil pH is too low.

- Symbiotic N fixation by legumes is greatly reduced. The symbiotic relationship requires a more narrow pH range for optimum growth than plants not requiring N fixation. The symbiotic bacteria for soybeans function best in a 6.0 to 6.2 pH range, while alfalfa bacteria function best in a 6.8 to 7.0 pH range.

- Highly acidic clay soils are less aggregated. This causes low permeability and aeration, an indirect effect because limed soils produce more crop residue. The residues contribute to better structure.

- Availability of nutrients such as P and Mo is reduced.

- The tendency for potassium (K) to leach is increased.

Aglime

When applied for optimum pH, aglime does much more than lower soil acidity and increase pH...

- **Aglime** reduces Al and other metal toxicities.
- **Aglime** improves the physical condition of the soil.
- **Aglime** stimulates microbial activity in the soil.
- **Aglime** increases CEC in variable charge soils.
- **Aglime** increases availability of several nutrients.
- **Aglime** supplies Ca and Mg for plants.
- **Aglime** improves symbiotic N fixation by legumes.

However, on tropical soils high in Fe and Al oxides, over-liming to pH values greater than 6.0 or 7.0 can seriously decrease yields, may cause soil structural deterioration, reduce phosphorus (P) availability and induce Zn, boron (B), and Mn deficiency.

Crop yields can rise sharply when acid soil is limed.

In this Tennessee study, first year soybean yields increased 6 bu/A and peaked at the recommended rate of 2 t/A of aglime. At least 500 lb/A was required to obtain a response to the aglime. Response to annual applications of aglime over a five-year period rose to about 14 bu/A with a net application of 4 t/A of aglime.

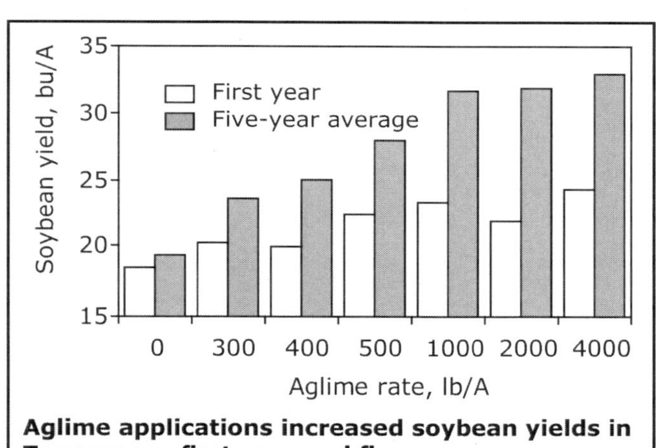

Aglime applications increased soybean yields in Tennessee—first year and five-year averages.

For best results, aglime is usually applied several months before planting to allow sufficient time to neutralize soil acidity. However, incorporating aglime immediately prior to planting can have major impacts on crop yields. In three Virginia studies, aglime increased soil pH within 16 weeks of application, resulting in yield increases of 100 bu/A in corn, 16 bu/A in barley, and more than 1 t/A in alfalfa.

Aglime incorporated immediately prior to planting increased crop yields in Virginia.

Lime, rate, t/A	Corn, bu/A	Barley, bu/A	Alfalfa, lb/A
	- - - - - - - Crop yield - - - - - - -		
0.0	21	49	303
0.5	87	—	—
1.0	105	61	1,229
2.0	104	56	1,674
3.0	—	56	1,817
4.0	110	65	2,262
6.0	121	60	2,262
8.0	—	59	2,369

Desirable pH Levels Vary

The soil pH range for optimum performance of the majority of field, forage, vegetable, fruit, and nut crops is often between 5.5 to 6.5 . Notable exceptions to this generalization include alfalfa, which performs best at a soil pH above 6.5, and blueberries, which perform better under more acidic (<pH 5.5) conditions. There are considerable geographic differences in desirable soil pH ranges, or ranges of crop tolerance to acid soil infertility. Some of these differences may be related to crop genetics and adaptation. Site-specific factors like aglime cost, specific soil micronutrient levels (from deficient to toxic), soil age (classification) and the potential for Al toxicity, soil test P levels, whether subsoil pH is higher or lower than in the surface soil, tillage system (no-till, plowed/disked, strip-tilled), and specific disease risks should all be considered in a soil pH maintenance and aglime program.

Soil properties differ in different areas. An ideal pH in one region may not be best in other regions. Optimum soil pH ranges among regions may exist for crops such as corn, soybeans, and alfalfa, but not for others, such as potatoes. Potatoes, soybeans, and other crops may be affected by diseases and/or micronutrient deficiencies if soil pH is below or above their individual requirements, regardless of geographic area. A working knowledge of the soil, as well as the crop, is essential to meet optimum soil pH and aglime requirements.

> **Review Questions:**
> 15. (T or F) Acid soils are bad for all crops.
> 16. Which of the following crops grows best in the pH range above 6.5: corn, alfalfa, wheat, soybeans, watermelons?

How Aglime Reduces Soil Acidity

The process and reactions by which aglime reduces soil acidity are complex. However, a simplified view will show how aglime works. As mentioned earlier, the pH of a soil is an expression of the H^+ activity. The principal source of H^+ in most soils with a pH below 5.5 is the reaction of Al with water, as shown by the following equation:

$$Al^{3+} + H_2O \longrightarrow Al(OH)^{2+} + H^+$$

This reaction releases H^+ (acidification), which in turn increases the amount of Al^{3+} ready to react again.

Aglime reduces soil acidity (increases pH) by converting some of the H^+ into water. Above pH 5.5, Al precipitates as $Al(OH)_3$. Thus, its toxic action and the main source of H^+ are eliminated.

The reaction works like this: While Ca^{2+} ions from the aglime replace Al^{3+} at the exchange sites, the carbonate ion (CO_3^{2-}) reacts in the soil solution, creating an excess of OH^- ions which in turn react with the H^+ (excess acidity), forming water. The overall process is illustrated in **Figure 2-3**.

Remember, the reverse of the above process can also occur. An acid soil can become more acid if a liming program is not followed. As basic ions such as Ca^{2+}, Mg^{2+}, and K^+ are removed, usually by crop uptake, they can be replaced by Al^{3+}. These basic ions can also be lost by leaching, again being replaced by Al^{3+}. This process will steadily increase H^+ activity, thus lowering soil pH, if the soil is not limed properly.

> **Review Question:**
> 17. Aglime increases soil pH by converting H^+ ions into __________ .

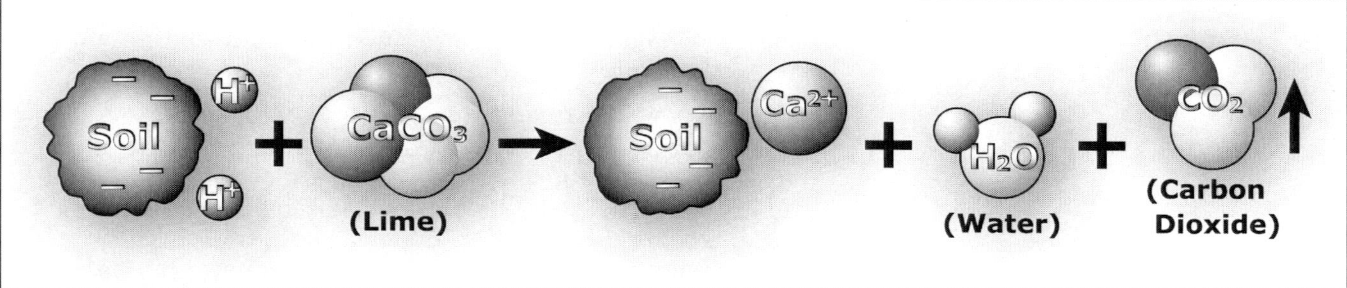

Figure 2-3. How lime reduces acidity.

Time and Frequency of Aglime Applications

For rotations that include legume crops, aglime should be applied three to six months before seeding, especially on very acid soils. Liming a few days before seeding alfalfa or clover, for example, can produce disappointing results because the aglime may not have time to react with the soil. If clover is to follow a fall seeded wheat crop, aglime should be applied at wheat planting. Regardless of the crop, aglime should be applied enough ahead of planting to allow reaction between the aglime and soil. Finely ground aglime can react very quickly if incorporated into the soil. Caustic forms of aglime...Ca oxide (CaO) and Ca hydroxide [Ca(OH)$_2$]...should be spread well before planting to prevent damage to germinating seeds.

General statements concerning the frequency of liming are probably unwise because so many factors are involved. The best way to determine reliming needs is to soil test. Samples should be taken at least once every three years...more often for sandy soils and for crops under irrigation. The following factors will influence frequency of liming.

- **Soil texture**–Sandy soils must be relimed more often than clay soils.

- **Rate of N fertilization**–High rates of ammonium (NH$_4{}^+$)–N generate considerable acidity.

- **Rate of crop removal**–Legumes remove more Ca and Mg than non-legumes. **Table 2-2** shows uptake of Ca, Mg, and S by several common crops.

Table 2-2. Calcium, Mg, and S taken up by some common crops.

Crop	Yield level	Pounds in total crop		
		Ca[1]	Mg	S
Alfalfa	8 tons	175	40	40
Coastal bermudagrass	8 tons	52	26	44
Corn	160 bu	39	52	27
Cotton	1,000 lb lint	14	23	20
Grain sorghum	8,000 lb	60	40	39
Oranges	540 cwt.	80	22	–[2]
Peanuts	4,000 lb	20	25	21
Rice	7,000 lb	20	14	21
Soybeans	60 bu	26	24	20
Tomatoes	40 tons	30	36	54
Wheat	60 bu	16	18	15

[1]Estimated [2]Not available

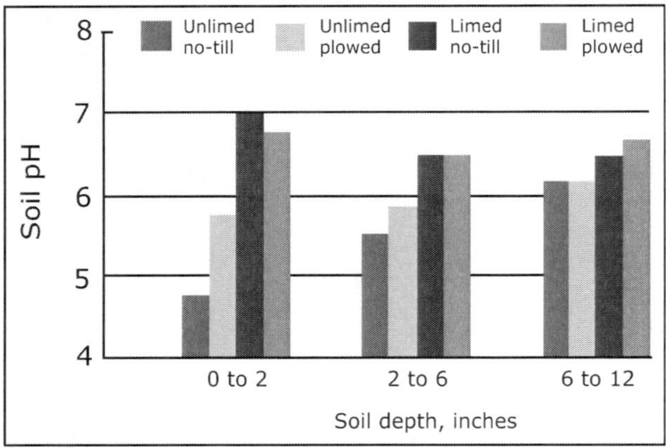

Figure 2-4. Soil pH at three depths after cropping with continuous corn for 10 years (150 lb N/A). (Kentucky)

Review Questions:
18. (T or F) Acid soils should be relimed every five years.
19. The best method of determining frequency of reliming is _____ _____.
20. (T or F) Soil texture and N fertilization both influence liming frequency.

• **Amount of lime applied**—Higher application rates usually mean the soil need not be relimed as often. But don't over-lime.

• **The pH range desired**—Maintenance of a high pH usually means that aglime must be applied more frequently than where an intermediate pH is satisfactory. Often the desired pH range is not reached because of under liming, poor quality aglime (coarse particles) or incomplete mixing. Soil tests can monitor pH changes with aglime.

• **Tillage system**—Acidity tends to develop more quickly in the upper two inches in no-till compared to other tillage systems. Aglime may need to be applied every three years or so to keep pH at desirable levels in the top few inches of soil, particularly when higher N rates are being applied (**Figure 2-4**).

• **Irrigated crops**—Soil pH should be monitored more frequently under irrigation because yields are usually higher. Soil nutrient status and pH can change quickly.

Selecting Liming Material— Quality Aspects

When selecting a liming material, check its neutralizing value, its degree of fineness, and its reactivity. Where soil Mg may be low or deficient, Mg content of the aglime should be a factor in selecting the material; that is, dolomitic aglime should be considered.

Neutralizing values of all liming materials are determined by comparing them to the neutralizing value of pure Ca carbonate ($CaCO_3$). Setting the neutralizing value of $CaCO_3$ at 100, a value for other materials can be assigned. This value is called the relative neutralizing value or $CaCO_3$ equivalent. The relative neutralizing values for several common liming materials are shown in **Table 2-3**.

Table 2-3. Relative neutralizing values of some liming materials.

Liming material	Relative neutralizing value, %
Calcium carbonate	100
Dolomitic lime	95-108
Calcitic lime	85-100
Baked oyster shells	80-90
Marl	50-90
Burned lime	150-175
Burned oyster shells	90-110
Hydrated lime	120-135
Basic slag	50-70
Wood ashes	40-80
Gypsum	None
By-products	Variable

When a given quantity of aglime is mixed with the soil, its reaction rate and degree of reactivity are affected by particle size. Coarse aglime particles react more slowly and less completely. Fine aglime particles react more rapidly and more completely.

Cost of lime increases with the fineness of grind. The goal should be to select a material that requires a minimum of grinding, yet contains enough fine material to cause a rapid pH change. As a result, agricultural liming materials contain both coarse and fine materials. Some states require a certain percentage of aglime to pass through certain mesh sizes. This guarantees the aglime will be of sufficient quality to neutralize acidity. The importance of particle size is shown in **Table 2-4**.

Table 2-4. The effect of fineness on availability of aglime.

	Years after application	
Mesh size	1	4
	- - - % reacted - - -	
coarser than 8	5	15
8 to 20	20	45
20 to 50	50	100
50 to 100	100	100

Table 2-4 tells a dramatic story about aglime particle size and degree of reactivity. Large particles (coarser than 8 mesh) were only 15% reactive. Smaller ones (80 to 100 mesh) reacted completely.

Although aglime reaction rate depends on particle size, initial pH, and degree of mixing with the soil, the chemical nature of the liming material itself is an important consideration. For example, CaO and Ca(OH)$_2$ react more quickly than CaCO$_3$. In fact, hydrated aglime can react so quickly it can partially sterilize the soil. If applied too near planting time, it can induce a temporary K deficiency because of the high Ca availability. Retarded plant growth and some crop death can result in extreme cases.

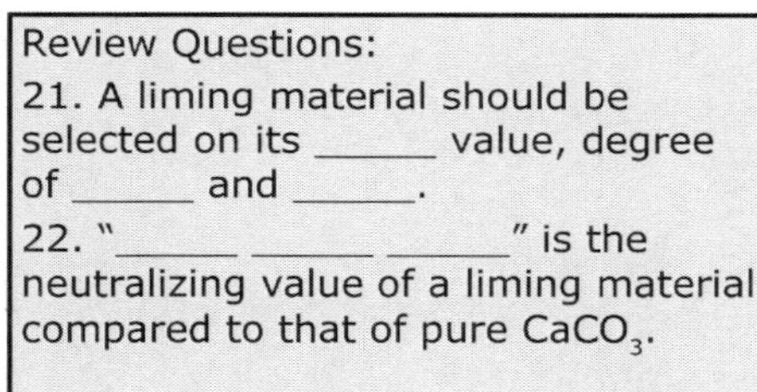

Lime Placement

Another important factor determining the effectiveness of aglime is placement. Placement for maximum contact with the soil in the tilled layer is essential. To begin with, most common liming materials are only slightly soluble in water, so distribution in the soil is a must for efficient aglime reaction. Even when properly mixed with the soil, aglime will have little effect on pH if the soil is dry. Moisture is essential for aglime-soil reaction to occur. When applying large amounts of aglime to clay soils, best mixing occurs when part is applied before primary tillage, the remainder being applied afterward. On sandy soils, one application incorporated by primary tillage will do.

In some cropping systems, such as perennial sods and no-till, mixing is only possible before seeding. Once the sod is established, the aglime must be topdressed. Surface-applied aglime reacts more slowly...and not as completely...as that mixed with the soil. So fields should be relimed before pH drops below the desired range to avoid excess acidity in the root zone.

Liming Materials

Although the common liming materials have been referred to in previous sections, short descriptions of CaO, Ca(OH)$_2$, calcitic lime (CaCO$_3$), dolomitic lime [CaMg(CO$_3$)$_2$], marl, stack dust, slags, and sludge follow:

- **Calcium oxide**—Also known as unslaked lime, burned lime, or quicklime, CaO is a caustic white powder, difficult to handle. It is manufactured by roasting CaCO$_3$ in an oven or furnace. Its purity depends on the purity of the raw material. When added to the soil, it reacts almost immediately, so when rapid results are required, it or Ca(OH)$_2$ is ideal. It should be mixed completely with the soil, because it rapidly cakes and can become ineffective.

- **Calcium hydroxide**—Frequently referred to as slaked lime, hydrated lime, or builders' lime, Ca(OH)$_2$ is a caustic, white, powdery substance, difficult and unpleasant to handle. Acid neutralization occurs rapidly when it is added to the soil. It is prepared by hydrating CaO.

- **Calcitic limestone and dolomitic limestone**—Deposits of high-grade ores of both sources are widespread throughout the U.S. They are most often mined by open-pit methods. Quality depends on the impurities, such as clay, they contain. Their neutralizing values (CaCO$_3$ equivalent) usually range from 65 to 70% to slightly more than 100%.

- **Marl**—Marls are soft, unconsolidated deposits of CaCO$_3$, occurring in many areas. Deposits may range in thickness up to 30 feet. They are mined by dragline or power shovel after the overburden has been removed. Marls are almost always low in Mg, and their liming value is inversely related to the amount of clay they contain.

- **Slags**—Several types of materials are classed as slags. Blast furnace slag is a by-product of the manufacture of pig iron. Basic slag is a product of the basic open-hearth method of making steel from pig iron. It is generally applied for its P content rather than for its value as a liming material. Electric-furnace slag results from the reduction (in an electric furnace) of phosphate rock in the preparation of elemental P. It is a waste product, usually marketed at a low price within a limited radius of production point.

- **Stack dust**—Stack dust is a by-product of cement production and contains a mixture of compounds including CaO, CaCO$_3$, K oxide (K$_2$O), K carbonate (K$_2$CO$_3$), and other materials. It is very finely divided and difficult to handle, but the small particle size makes it an ideal product for use in fluid suspensions. Presence of oxides, however, can cause pH of these suspensions to be around 12.

- **Water softening sludge**—This material is a byproduct of municipal water softening plants or industrial boiler feedwater treatments. It is largely $CaCO_3$, very finely divided, usually wet, and often discarded as a waste. When available, it is an excellent source of material for the production of fluid lime suspensions.

Crop Response to Liming Acid Soils

Production Concept 2-1 showed how important lime can be to crop production. There is no doubt that excess soil acidity limits crop yields. It can also cut fertilizer use efficiency by as much as 50%. Untreated soil acidity impacts both crop yields and farmer profits.

Table 2-5 illustrates how dramatically aglime can affect yields. For corn, response to 6.0 tons of aglime per acre was 100 bushels. First cut alfalfa response was nearly a ton per acre at the 6.0 ton aglime rate.

Table 2-5. Response of corn and first cut alfalfa to aglime applied just before planting.

Rate, tons/A	Corn response, bu/A	First cut alfalfa response, lb/A
0.0	–	–
1.0	84	926
4.0	89	1,888
6.0	100	1,959

Liming benefits yields in both conventional and no-till systems. In Pennsylvania studies, it was shown that surface applications of aglime in no-till primarily affected the surface 2 inches of soil. However, this pH change benefited crop yields and herbicide performance. For example, wheat yields were increased by as much as 17 bushels per acre, as shown in **Table 2-6**.

Table 2-6. Wheat yield response due to liming.

Rate, lb/A	Wheat yield, bu/A
0	52
3,000	69
6,000	71
9,000	69

High pH Soils: Calcareous, Saline, and Alkali (Sodic)

Many soils in arid climates have high pHs which can affect their properties and influence productivity. Obviously, they do not require aglime, but their high pHs do affect nutrient availability, soil fertility, and fertilizer management.

- **Calcareous soils contain** free lime (undissolved $CaCO_3$) with pHs generally ranging between 7.3-8.4. The presence of free lime does influence some management practices such as herbicide use, P placement (because of fixation), and micronutrient availability, particularly Fe. Lowering the pH of calcareous soils is not usually economical. With proper management, these are some of our most productive soils.

- **Saline soils** contain salts in quantities high enough to limit crop growth because plants cannot take up sufficient water to function properly. Plants growing on saline soils often exhibit wilting, even when soil water content is adequate. The degree of salinity is measured in the soil testing laboratory as electrical conductivity (EC). Saline soils can be reclaimed by leaching salts out of the root zone with high quality water. Since crops differ in their salt tolerance, a best management practice is to select those crops which are known to be salt tolerant. **Table 2-7** compares some common crops with regard to their tolerance to salinity. These soils usually have pHs lower than 8.5.

Table 2-7. Tolerance of some common crops to soil salinity.

Tolerance level		
Good	Moderate	Poor
Barley	Cereal grains	Most clovers
Sugar beet	Corn	Field bean
Canola	Alfalfa	Celery
Cotton	Orchardgrass	Apple
Bermudagrass	Rye	Orange
Sweet clover	Peach	
	Soybeans	

- **Sodic (alkali) soils** contain excessive amounts of sodium (Na) on the soil CEC sites. Soils are usually classified as alkali if Na saturation exceeds 15% of the CEC. They usually have pH values of 8.5 and above. Excess Na disperses the soil, limiting movement of air and water because of poor physical properties. Water tends to pond on alkali soils.

Such soils can be reclaimed by replacing the Na on the CEC complex with Ca, gypsum application being the most common treatment. However, elemental S can also be used if the soil is calcareous. The acidity generated from oxidation of the elemental S helps dissolve $CaCO_3$, making the Ca available to displace Na. Successful reclamation requires that the Na be leached out of the root zone, and poor water movement can make that task difficult. Deep ripping and/or manure application has been used to improve internal water movement.

Sometimes sodic soils will also be saline. Saline/sodic soils are typically characterized by a Na saturation greater than 15% of the CEC, high EC, and a pH of 8.4 or less. Their reclamation is the same as that for sodic soils. ⟨SFM⟩

Nitrogen

Nitrogen (N) is essential for plant growth and is a part of every living cell. It plays many roles in plants and is necessary for chlorophyll synthesis. Symptoms of N deficiency in plants usually include chlorosis or yellowing. Adequate N is important for water use efficiency. The amount of soil N in available form (inorganic) is very small, while a high proportion is unavailable (organic) N. Nitrates are readily available for crops, mobile in the soil, and can be lost through denitrification. Use of nitrification inhibitors or slow release forms of N can improve N use efficiency. Nitrogen fixation must occur before N can be used by plants. Harvested crops remove considerable amounts of N from the soil. Most forms of N tend to increase soil acidity. Many different N fertilizers are available, containing various forms and contents of N.

An Essential Plant Nutrient

Nitrogen is essential for plant growth. It is a part of every living cell. Plants require large amounts of N for normal growth, **Table 3-1**.

Table 3-1. Crops have high N requirements.

Crop	Yield level		Pounds of N taken up in total crop
Alfalfa[1]	8	tons	450
Coastal bermudagrass	8	tons	368
Corn	160	bu	213
Cotton	1,500	lb (lint)	180
Grain sorghum	7,500	lb	222
Oranges	540	cwt	265
Peanuts[1]	4,000	lb	240
Rice	7,000	lb	112
Soybeans[1]	60	bu	315
Tomatoes	40	tons	232
Wheat	60	bu	113

[1]Legumes get most of their N from the air.

Plants take up most of their N as the ammonium (NH_4^+) or nitrate (NO_3^-) ion. Some direct absorption of urea can occur through the leaves, and small amounts of N are obtained from materials such as water soluble amino acids. Except for rice, most agronomic crops take up most of their N as NO_3^-.

Nitrate is very mobile in the soil (compared to NH_4^+) and moves with soil water to root surfaces where it can be absorbed. Also, with time and under proper soil conditions, soil organisms convert all N forms to NO_3^-. Research has shown that crops use substantial amounts of NH_4^+ if it is present in the soil. Certain corn hybrids have a high requirement for NH_4^+, and that helps boost yields. Wheat has also shown benefits of NH_4^+ nutrition. One reason for the higher yields is that NO_3^- reduction in the plant requires energy (NO_3^- is reduced to NH_4^+, then converted to amino acids inside the plant). The energy is supplied by carbohydrates which could otherwise be used in grain formation.

Nitrogen is necessary for chlorophyll synthesis and, as a part of the chlorophyll molecule, is involved in photosynthesis. Lack of N and chlorophyll means the crop will not utilize sunlight as an energy source to carry on essential functions such as nutrient uptake. Nitrogen is utilized by plants to synthesize amino acids. It is a component of vitamins and energy systems in the plant, as well as amino acids, which form plant proteins. Thus, N is directly responsible for increasing protein content. **See Production Concept 3-1**.

Plant Deficiency Symptoms

Adequate N produces a dark green color in the leaves, caused by a high concentration of chlorophyll. Nitrogen deficiency results in chlorosis (a yellowing) of the leaves because of declining chlorophyll. This yellowing starts first on oldest leaves, then develops on younger ones as the deficiency becomes more severe. Death (necrosis) of leaf tips and margins can occur in severe cases, beginning on mature leaves first. See photos of deficiency symptoms in Appendix A.

Green pigment in chlorophyll absorbs light energy needed to initiate photosynthesis. Chlorophyll helps convert carbon (C), hydrogen (H), and oxygen (O) into simple sugars. These sugars and their conversion products stimulate plant growth and development.

Slow growth and stunted plants are also indications of N deficiency. Small grains and other grass-type plants tiller less when N is in short supply.

Inadequate N leads to low protein in seed and plant vegetative parts. Deficient plants usually have fewer leaves, and certain crops such as cotton may reach maturity earlier than plants with adequate N. Corn properly fertilized with N will have a lower moisture content in the grain at harvest than corn supplied with insufficient N. Nitrogen is sometimes blamed for delayed maturity. Excess N can increase vegetative growth, reduce fruit set, and adversely affect quality. Delayed maturity, however, is usually caused by deficiencies of other nutrients, not by too much N.

Soil Fertility Manual

Nitrogen — Builder of Protein

The world needs more than 525 million pounds of protein per day.

The world needs a lot of protein. Sound agronomic practices, including adequate use of N fertilizer, help to produce that protein requirement in the form of higher, more protein-rich crop yields.

Assuming an average of 40 grams (about 1.4 ounces) of protein per person per day for the entire world population, annual consumption would be around 100 million tons of protein. This protein reaches dinner tables from plants directly or via animals, birds, or fish that have consumed plant protein.

We live in a protein hungry world, and an important key to protein production is N fertilization. Proteins are polymers of amino acids linked together into chains. An amino acid is a small molecule that acts as a building block of every cell. Essentially all the proteins on earth are made from 20 different amino acids. Each amino acid contains a central carbon (C) bonded to a hydrogen (H) atom, a caboxyl group (COOH), an amino group (NH_2), and a unique side chain.

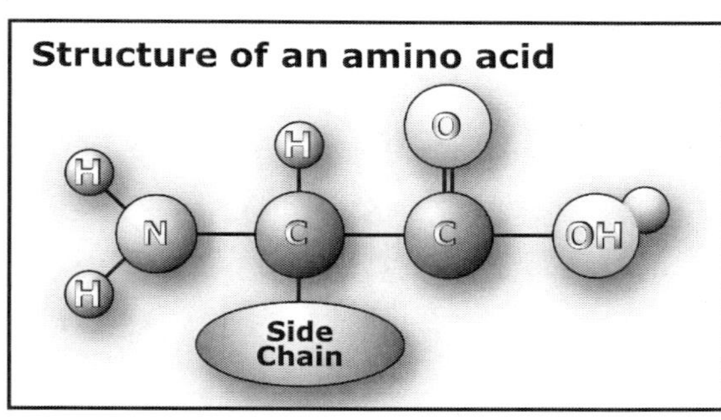

Structure of an amino acid

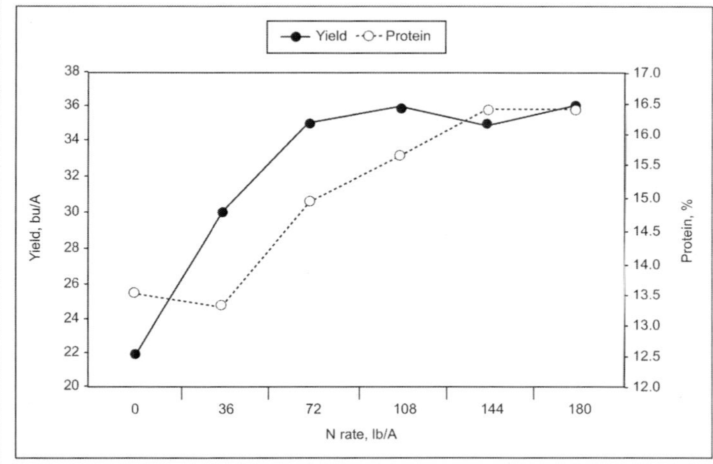

Nitrogen is essential for crops to achieve optimum yields. And because it is also a critical component of each amino acid in each protein, N increases protein content of plants directly.

A bushel of hard red winter wheat produces about 70 loaves of bread. For high quality bread, the wheat must have high protein and that requires a good supply of available N. Adequate P and K improve the plant's use of N to produce more protein and improve product quality.

Anytime a missing nutrient increases yield when it is applied, WUE is increased. **Table 3-2** shows how N more than doubled corn production from the same amount of water on a sandy loam soil. Increasing N from 100 to 200 lb/A added 66 bu/A to the yield. Applying the 200 lb N rate in eight 25 lb increments rather than one added 34 bu/A.

Table 3-2. **More N at the right time doubles corn yield from the same amount of water.**

N rate, lb/A	Application method/timing	Yield, bu/A	Yield response, bu/lb N
0		43	
100	All at planting	92	0.49
	Four 25 lb applic. (through irrigation)	154	1.11
200	All at planting	158	0.58
	Eight 25 lb applic. (through irrigation)	192	0.75

Minnesota

Nitrogen increases bushels of corn per inch of available water regardless of the amount of water. In **Table 3-3**, 150 lb N/A gave a little over 2 bu more yield per inch of water with wet conditions, almost 2.75 bu more per inch of water in the dry year.

Table 3-3. **Nitrogen increased corn yield in wet or dry weather.**

N rate lb/A	Grain yield, bu/A		WUE, bu/in. H_2O	
	Dry	Wet	Dry	Wet
0	75	96	4.82	4.68
150	114	152	7.48	6.98
N response, bu/A:	39	56	—	—

Colorado

In Arizona, N added up to 100 lb more wheat per inch of water. In Texas, N helped grow 160 lb more sorghum per inch of water used **(Table 3-4)**.

Table 3-4. **Nitrogen produced more grain sorghum per inch of water, three-year average.**

N rate, lb/A	Yield, lb/A	WUE, lb grain/in. H_2O
0	4,530	190
120	6,940	329
240	7,250	349

Texas

Applying optimum rates of N and other fertilizer nutrients... not too much nor too little...to meet crop needs results in improved yield and WUE while minimizing potentially negative effects on the environment.

Fertilizer and Optimum Soil Fertility Stretch Moisture

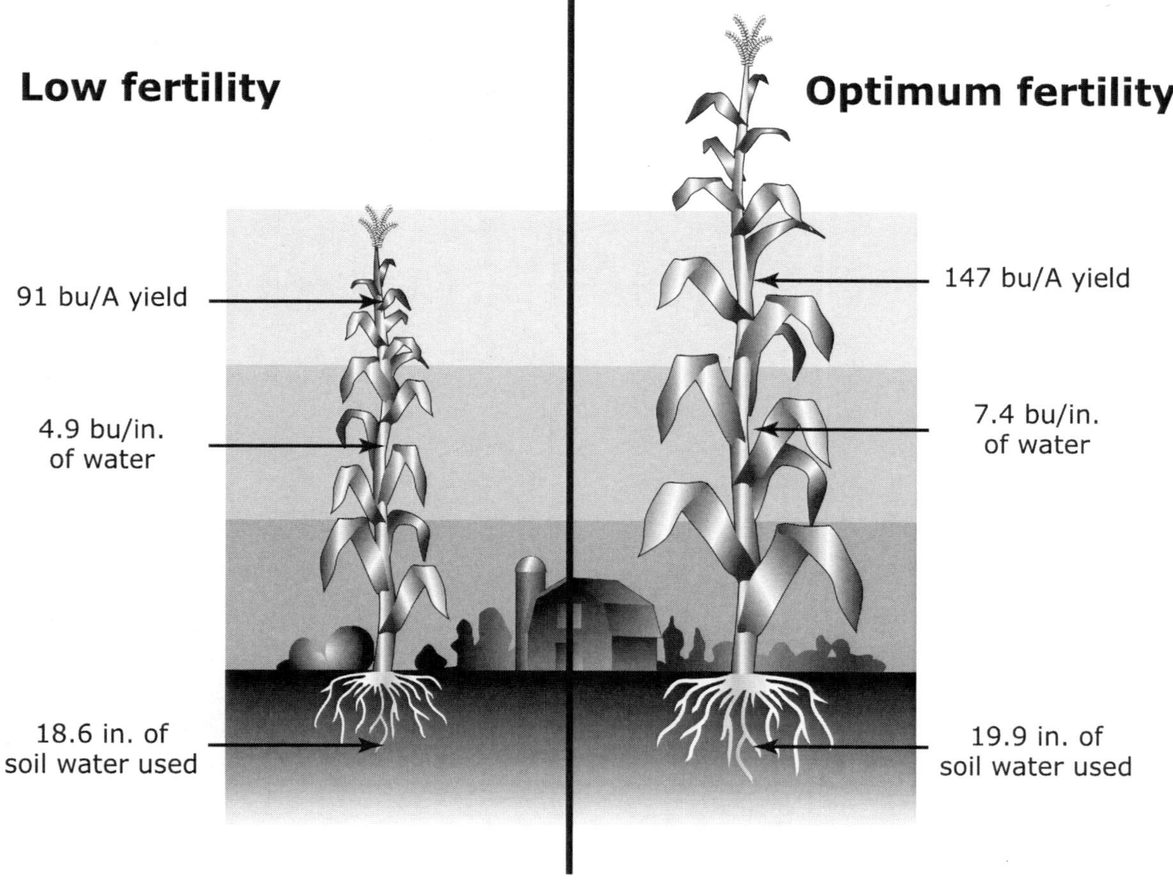

The next best thing to rain…that is what fertilizer has been called. It deserves the praise. Why? Because:

- Fertilizer helps produce more crop per inch of water, as shown above;
- Fertilizer sends roots deeper to find moisture in the subsoil;
- Deeper, more dense root systems take up more nutrients and moisture;
- Fertilizer creates a larger, thicker canopy fast to reduce soil water evaporation;
- Good vegetative cover slows runoff and enables soil to soak up water;
- Fertilizer helps crops start fast to shade out moisture-robbing weeds.

The greatest yield increases from fertilizers, percentage-wise, frequently occur in years of moisture stress.

Soil N Transformations

By understanding the transformations of N and other nutrients that take place in the soil, the efficiency of their management can be increased. All nutrient sources, inorganic and organic, can be utilized in such a way that crop yields, farmer profits, and environmental protection can be improved.

Transformations are the result of various soil reactions which determine where nutrients can be found in the soil. The entire set of reactions for a given nutrient is referred to as its cycle, appropriate terminology because nutrients are not destroyed during the various reactions taking place. They are simply relocated or converted to another form. The N cycle is shown in **Figure 3-1**.

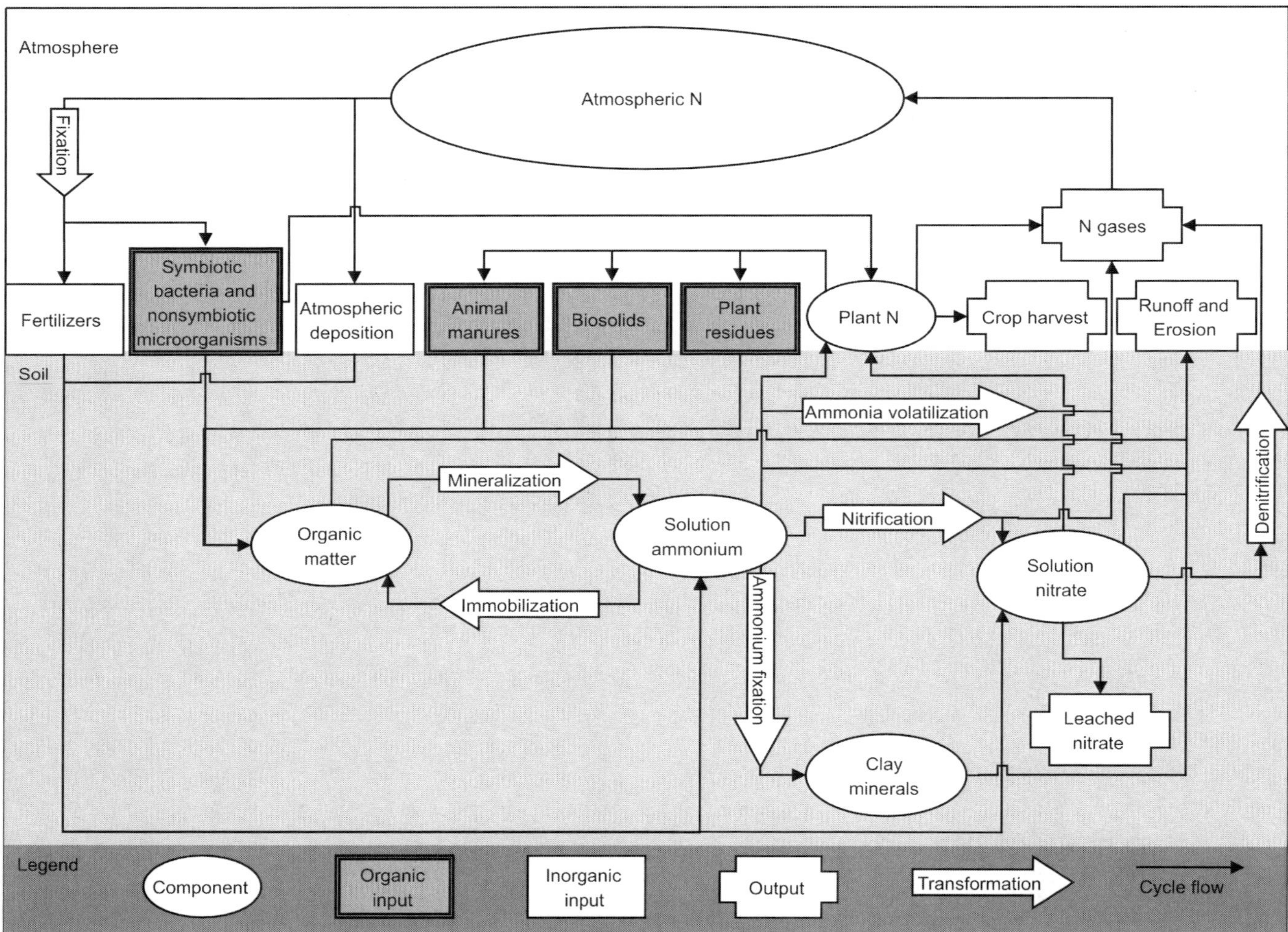

Figure 3-1. The nitrogen cycle.

Nitrogen in the Soil and Air

The atmosphere is about 78% N. Each acre of the Earth's surface is covered by about 37,000 tons (75 million lb), but this N is an inert gas (N_2). It must be changed chemically before plants can use it. Unlike the atmosphere, which contains a virtually unlimited supply of N, the amount of soil N (in plant-available form) is small. Very little N is found in the rocks and minerals from which the soil was formed. Most comes from the atmosphere.

Review Question:

9. (T or F) Most soil rocks and minerals contain N.

10. Most all of the soil's N comes from the _____.

11. Each acre of the Earth's surface is covered with about _____ tons of N.

Soil N occurs in two major forms: organic N (contained in organic matter) and inorganic N. The two primary forms of inorganic soil N are NH_4^+ and NO_3^-. Ammonium is attracted to and held by the surfaces and edges of negatively charged soil particles, while NO_3^- moves freely in soil water. Ammonium is also found in the soil solution.

Mineralization and Immobilization of Nitrogen

The soil contains a relatively large proportion of unavailable (organic) N and a small proportion of available (inorganic) N, illustrated in **Figure 3-2**.

Organic N may represent 97 to 98% of the total N in the soil. Inorganic N usually represents only 2 to 3%. So the process by which unavailable organic forms are converted to available forms is important to plant growth. This process is called **mineralization**. It occurs as microorganisms decompose organic materials for their energy supply. As the organic matter is decomposed, the organisms use some of the energy released plus part of the essential nutrients in the organic matter. When the organisms have used all the nutrients they need, excess amounts (such as N) are released into the soil for plant growth.

Nitrogen can also be converted from inorganic to organic forms, as the double arrows show. This process is called **immobilization.** It is the reverse of mineralization. Immobilization occurs when crop residues high in carbon (C) and low in N content are incorporated into the soil.

As microorganisms vigorously decompose the new energy supply in these crop residues, they need N to build protein for their body tissues. Unless residues are relatively high in N, organisms take up inorganic N from the soil to support their growth. So the inorganic N in the soil is converted into organic N in microbial proteins, unavailable for plant growth. But much of this N is gradually returned to the available form as the bacterial bodies decompose.

Figure 3-2. Most soil N is contained in the organic pool and is not readily available for plant use.

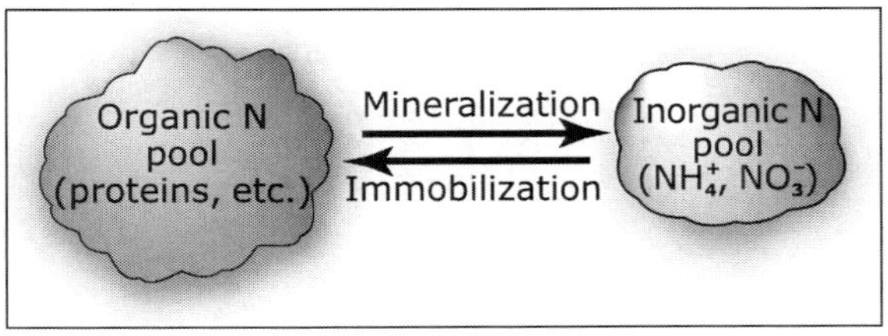

Mineralization and immobilization occur simultaneously in soils. Whether the soil shifts toward an organic or inorganic N pool depends largely on the C:N ratio of the decomposing organic materials. Materials with wide C:N ratios (above 30:1) favor immobilization. Materials with narrow C:N ratios (less than 20:1) favor more rapid mineralization. At C:N ratios in the range of 20 to 30:1, the two processes are about equal. **Table 3-5** shows C:N ratios of selected organic materials.

Table 3-5. Ratios of C:N for selected organic materials.

Material	C:N ratio
Undisturbed top soil	10:1
Alfalfa	13:1
Rotted barnyard manure	20:1
Corn stalks	60:1
Small grain straw	80:1
Coal and shale oil	124:1
Oak	200:1
Spruce	1000:1

Review Questions:

12. The most plant-available forms of soil N are _____ and _____ ions of soluble N compounds. The least available form is organic N.

13. The process by which unavailable organic N forms are converted to available N forms is known as _____.

14. (T or F) Materials with wide C:N ratios cause immobilization to dominate over mineralization.

15. (T or F) Alfalfa has a wider C:N ratio than does undisturbed top soil.

When immobilization of soil N exceeds mineralization, there may be practically no N available for growing crops unless N fertilizers have been applied in a band near the roots. This is called the NO_3^- depression period. It is a critical time for crops. How long it lasts depends on three factors: 1) the C:N ratio of the decomposing material, 2) the quantity of crop residue added to the soil, and 3) the environmental conditions in the soil. Adding more crop residue generally lengthens the period. Supplying adequate N generally shortens it. To avoid the problem or to minimize its impact, residues should be incorporated well ahead of planting to allow early decomposition, or left on the soil surface in no-till production.

Nitrification and Denitrification

An early product of organic matter decomposition is NH_4^+, resulting from the breakdown of proteins to amino acids (aminization). The conversion from amino acids to NH_4^+ is called ammonification. Under conditions favoring plant growth, much of the NH_4^+–N in soils will be converted to NO_3^-–N by certain nitrifying bacteria. This process is called nitrification (**Figure 3-3**). It is important for several reasons:

- Nitrate is readily available for use by crops and microorganisms. Organisms also use NH_4^+ under aerated conditions.
- Nitrate is highly mobile in the soil. It moves freely with soil water. Much more NO_3^-–N may leach through the soil profile—more on deep, sandy soils than on fine-textured soils with moderate drainage and high rainfall. Nitrogen management, though, can control leaching into groundwater.
- Nitrate can be lost through denitrification, a process whereby NO_3^- is reduced to gaseous nitrous oxide (N_2O) or N_2 and lost to the atmosphere.

Denitrification usually occurs in soils high in organic matter, under extended periods of waterlogged conditions (absence of

Figure 3-3. Ammonium reacts with O_2 in the presence of nitrifying bacteria to produce NO_3^-. Hydrogen ions are also released, increasing soil acidity.

$$\underset{\text{Ammonium}}{2\,NH_4^+} + \underset{\text{Oxygen}}{3\,O_2} \xrightarrow[\text{Nitrifying bacteria}]{\text{Nitrification}} \underset{\text{Nitrate}}{2\,NO_3^-} + 2\,H_2O + \underset{\text{Hydrogen}}{4\,H^+}$$

O_2) and as temperature rises. Following are descriptions of soil conditions which have the greatest influence on nitrification and denitrification.

- **Soil pH**—Nitrification rates are usually low in acid soils. Nitrification has occurred in the pH range of 4.5 to 10.0, but a pH of about 8.5 is optimum for nitrification. Liming strongly acid soils benefits nitrifying bacteria. Liming has been shown to increase denitrification under some conditions.

- **Moisture**—Nitrifying bacteria remain active in very dry conditions, but inactive in waterlogged soil. Soils with sufficient moisture to grow crops will have enough moisture for normal nitrification. Wet soils do not contain enough O_2 to supply nitrifying bacteria. As a result, very little NO_3^- will be produced. When O_2 is excluded from the soil, bacterial denitrification can occur. This can reduce N availability sharply because of increased N loss.

- **Temperature**—Nitrification begins slowly...just above freezing. It continues to increase as soil temperature increases, up to above 85°F, at which point rates decrease. Denitrification reaction rates also increase with increasing soil temperatures.

- **Aeration**—Nitrification requires O_2. Well-aerated soils of medium to coarse textures have high O_2 contents and support rapid nitrification because of good drainage and air movement between the soil and the aboveground atmosphere.

- **Plant residues**—Denitrification occurs as soil bacteria break down organic residues. Larger amounts of residues combined with low O_2 supplies in the soil enhance denitrification reactions and N losses.

Nitrogen Fixation

When atmospheric nitrogen (N_2) combines with H_2 or O_2, a process called fixation occurs. This process must occur before the N can be used by plants. Fixation can occur in different ways.

Biological fixation may be symbiotic or non-symbiotic. Symbiotic N fixation refers to bacteria fixing N while growing in association with a host plant. It benefits both...the microorganisms and the host plant. The most widely known example is the association between Rhizobium bacteria and legume roots. Bacteria form nodules on the roots. Bacteria in these nodules fix N from the atmosphere and make it available to the legume. The legume furnishes carbohydrates which give bacteria the energy to fix N. How much N do legume bacteria fix? Estimates range from just a few pounds to over 500 lb/A/yr. Some commonly accepted values are shown in **Table 3-6**. Symbiotic N fixation by legume bacteria is considered the most important natural N source in soils. Research is now looking for N-fixing organisms that will grow and fix N with grasses.

Table 3-6. Estimated annual N fixation by various legume crops.

Legume	N fixed, lb/A/yr
Alfalfa	195
Ladino clover	180
White clover	105
Soybeans	100
Peas	95
Lentils	70
Lespedezas (annual)	85
Peanuts	40

It should be noted that legumes are net removers of N. That is, they take up and use more than is fixed. Numbers shown are averages from various sources.

Virginia research shows that P and K affect nodulation and hence N fixation. Note how P and K increased soybean nodule number, % N in nodule, and seed protein production (**Table 3-7**).

Non-symbiotic N fixation is carried out by free-living bacteria in the soil. The amount of N fixed by these organisms is much

Table 3-7. Phosphorus and K increase soybean yields, nodulation, and total seed protein production.

Annual rate, lb/A P_2O_5	K_2O	2 yr avg. yield, bu/A	Nodule number/ plant	Nodule green wt., lb/A[1]	N in nodule, %	Seed protein 2 yr avg., %	Seed protein production 2 yr avg., lb/A
0	0	25.5	35	285	3.19	41.8	639
120	0	26.4	59	525	3.92	41.8	662
0	120	46.8	79	745	3.37	39.2	1,096
120	120	54.8	114	1,406	3.61	39.2	1,289

[1]Assuming a depth of 6.75 inches.

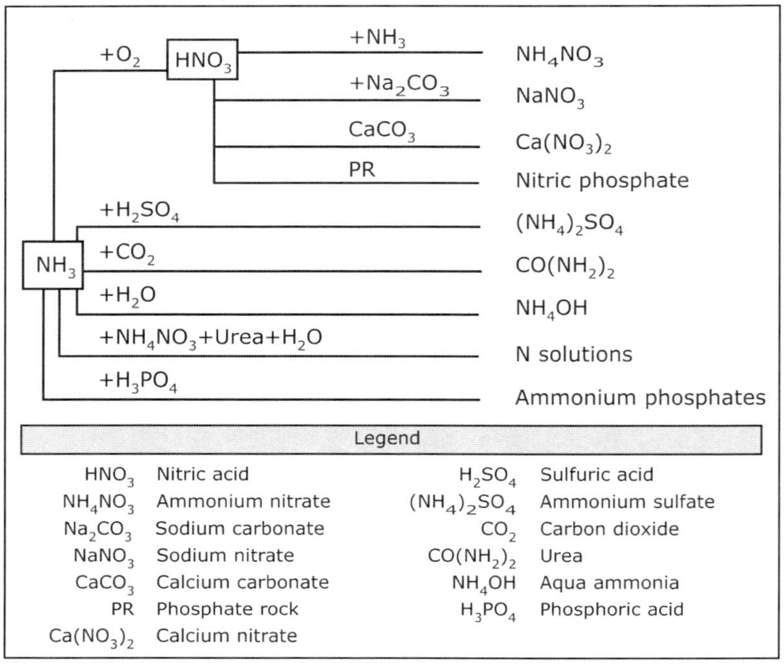

Figure 3-4. Ammonia (NH₃) is the basic product from which other N sources are produced.

less than the amount fixed symbiotically. Most estimates indicate that a maximum of about 20 lb N/A is fixed annually by free-living bacteria.

A small amount of N is added to the soil by natural oxidation. Heat generated by lightning causes O_2 to react with N_2 in the air, eventually forming $NO_3^- - N$. Rain and snow add only about 5 to 10 lb of N/A/yr from this source. In addition, air pollution deposits some N. Ammonium ion deposition ranges from 0.1 to 5 lb/A/yr, while NO_3^- deposition contributes from 0.2 to 6 lb of N/yr.

Industrial processes fix N_2 very effectively in forms available to plants. The most important process synthesizes ammonia (NH_3) from N_2 and H_2, as follows:

$$N_2 \quad + \quad 3H_2 \quad \xrightarrow{\text{Heat; pressure; catalyst}} \quad 2NH_3 \text{ (Anhydrous ammonia)}$$

The H_2 is usually obtained from natural gas. The N_2 comes directly from the air. **Figure 3-4** shows how NH_3 is then used to make other fertilizer materials.

Losses of Nitrogen

Harvested crops remove much N from the soil. The amount depends on kind and quantity of crop. Although removal in crops is not usually considered a loss, in reality it is. The net effect of crop removal is lower soil N levels. Other kinds of N losses are described below.

- **Ammonium reactions**—When N fertilizer materials such as NH_4NO_3 or $(NH_4)_2SO_4$ are applied to the surface of alkaline or calcareous soils, a chemical reaction may cause the loss of N as NH_3 gas, a process called volatilization. Similar reactions can occur on recently limed soils. Volatilization losses can be high under some high temperature and certain moisture conditions. To avoid such loss, fertilizer applied to alkaline or calcareous soils should be incorporated.

- **Urea**—Urea N applied to the soil surface converts rapidly to NH_3 or NH_4^+ with adequate moisture, proper temperature, and the presence of the enzyme, urease. This NH_3 can be lost to the atmosphere through volatilization. Urea loss can be avoided by incorporating, applying when temperatures are low, or irrigating immediately to carry the urea into the soil.

- **Anhydrous ammonia**—Anhydrous NH_3 is a gas when not under pressure. It must be placed below the soil surface to prevent volatilization loss. Loss can occur when NH_3 is applied to extremely moist soil. Ideal application time is when moisture is below field capacity—moist, but not waterlogged nor too dry. Sandy, low cation exchange capacity (CEC) soils need deeper NH_3 application than clay-type soils.

All the major commercially produced N fertilizers are highly soluble when applied to the soil. Organic sources such as animal manures, crop residues, and cover crops ultimately release soluble N as they decompose. Left unused by a growing crop, all sources of nutrients are eventually converted to NO_3^--N. In the NO_3^- form, N is subject to loss through surface erosion, leaching, and denitrification.

As NH_4^+, N is stable in the soil, being held on CEC sites by clays and organic matter. There are good reasons for keeping N in this form, at least until near the time the crop needs it.

- Ammonium-N is not subject to leaching, so the potential for movement into groundwater is minimized or eliminated.

- Some corn hybrids, wheat, cotton, and other crops produce higher yields when fed a mixture of NH_4^+ and NO_3^-.

- Soil N is not subject to denitrification when in the NH_4^+ form.

An important part of fertilizer N management is to apply proper rates and sources, place the N for best use efficiency, and time applications to the periods of greatest crop need. Sometimes it is difficult or impossible to achieve all these goals. However, through the use of nitrification inhibitors or by using slow release forms of N, use efficiency can be increased significantly.

- **Nitrification inhibitors**—These products block the conversion of NH_4^+ to NO_3^- by deactivating nitrifying bacteria for varying periods of time, sometimes up to three months. Results are variable, but yield increases of as much as 50% have been achieved when inhibitors are used properly. The greatest potential benefit from the use of nitrification inhibitors is with fall or early spring applied N, on sandy soils, on poorly drained soils, and where high rainfall is likely.

- **Slow release nitrogen**—Urea-formaldehyde fertilizers are made by reacting urea with formaldehyde to form compounds only slightly water soluble. High cost often prohibits their use on field crops. Main uses are on lawns, golf courses, and other specialty crops. Sulfur-coated urea is another type of slow release N fertilizer.

- **Controlled release urea (CRU)**—These products are made by coating urea granules with synthetic polymers. The polymer used and the thickness of the coating influence the number of days after application before N release into the soil. The range of CRU products available allows their use on high value crops as well as crops such as corn and wheat.

Review Question:

29. Nitrification inhibitors work by deactivating the bacteria which convert _______ to _______.

How Nitrogen Affects Soil Acidity

When the nitrification process converts the NH_4 to NO_3^-, H^+ ions are released (**Figure 3-3**). This is a source of soil acidity. So N sources (fertilizers, manure, legumes) containing or forming NH_4^+–N increase soil acidity unless the plant absorbs the NH_4^+ ions directly. Also, NO_3^- is a major factor associated with leaching of such basic ions as calcium (Ca^{2+}), magnesium (Mg^{2+}), and K^+ from the soil. The NO_3^- and bases move out together. As these bases are removed and replaced by H^+, soils become more acid.

When the mineralization process decomposes soil organic matter, the first N product is NH_4^+. As it converts to NO_3^-, H^+ ions are released. This, like inorganic NH_4^+ fertilizers, causes soil acidification. Such N carriers as $NaNO_3$ and $Ca(NO_3)_2$ leave basic cations...Ca^{2+} and sodium (Na^+)...in the soil. This makes soil less acidic. **Table 3-8** shows how different N sources affect the acidity or basicity of soils.

Table 3-8. Effects of nitrogen sources on acidity or basicity of soils.

N source	Chemical formula	N, %	lb $CaCO_3$/lb of N[1]
Ammonium sulfate	$(NH_4)_2SO_4$	21	5.2
Anhydrous ammonia	NH_3	82	1.8
Ammonium nitrate	NH_4NO_3	34	1.8
Urea	$CO(NH_2)_2$	46	1.8
Urea-ammonium nitrate solution (UAN)	$CO(NH_2)_2+NH_4NO_3$	28-32	1.8
Monoammonium phosphate (MAP)	$NH_4H_2PO_4$	10	5.4
Diammonium phosphate (DAP)	$(NH_4)_2HPO_4$	18	3.6
Potassium nitrate	KNO_3	13	-2.0(B)

[1]Amount of pure $CaCO_3$ required to either offset the acid-forming reactions of 1 lb of N or the amount of $CaCO_3$ required to equal the acid-neutralizing effects of 1 lb of N (B). Most of the acid-forming effects are due to the activities of soil bacteria during nitrification.

Nitrogen Sources

Organic matter decomposition provides more than 90% of the N derived from the soil. Many soils contain little organic matter, usually 2% or less. Soil organic matter contains about 5% N, but only about 2% of the organic matter is decomposed each year, often less. Each 1% of organic matter releases an estimated 10 to 40 lb of N each year, not enough to meet the needs of most agronomic crops. Further, the rate of release is affected by management practices. Conservation tillage, being practiced on more cropland each year, results in cooler soils, slower rates of organic matter decomposition, and lower rates of N release.

At one time, nearly all N fertilizer used came from organic materials. Peruvian guano (seabird droppings) was the first N fertilizer sold in the U.S. Such materials as chicken litter, cottonseed meal, and steamed bonemeal have also been used. Some of these are still used in specialized cases. But less than 0.5% of the fertilizer sold in the U.S. today is organic. Most N fertilizer comes from the commercial fixation of atmospheric N into NH_3 and further processing of NH_3 into other compounds.

- **Anhydrous Ammonia**—Anhydrous NH_3 contains more N than any other N fertilizer (82%). It is stored under pressure as a liquid. It is applied to the soil from high-pressure tanks by injection through tubes running down the rear of in-soil banding equipment or by metering into irrigation water for flood or furrow irrigation (not sprinkler irrigation). Applying anhydrous NH_3 can be difficult on rocky, wet or dry, cloddy soils. On sod crops, it can cause temporary injury to roots and results in rough surface conditions.

  A gas at normal temperatures and atmospheric pressure, some NH_3 can be lost during and after application. Soil physical and chemical conditions affect the amount of loss. Soil moisture content, depth of application, spacing of application bands or release points (on a sweep), and soil CEC affect how well the soil retains NH_3. Low CEC soils such as sands may require deeper NH_3 application to avoid volatilization losses.

  If a soil is very dry and cloddy during NH_3 application, the slit behind the applicator knife will not seal, and NH_3 may volatilize. Soil moisture content near field capacity is ideal for NH_3 retention. Water-logged conditions also increase loss possibilities because of difficulty in sealing behind the knives.

  Narrow applicator knife spacings tend to reduce NH_3 losses because of reduced NH_3 concentrations at the point of injection.

  Lower rates of application also reduce NH_3 concentrations at release points and reduce possibilities of loss.

  Low pressure application of NH_3 (Cold-Flo®) results in NH_3 release as a liquid below the soil surface, lowering the possibility of volatilization losses. Less soil cover is then required to retain the NH_3.

  Everything considered, volatilization losses of NH_3 are usually quite low and are not a large economic factor.

- **Aqua NH_3 and solutions**—Aqua NH_3 is made by dissolving NH_3 in water. It has properties similar to anhydrous NH_3 and must be placed beneath the soil surface to prevent NH_3 loss. Nitrogen solutions are manufactured by blending concentrated liquors of NH_4NO_3, urea, and sometimes aqua NH_3. Nitrogen solutions are sometimes produced by dissolving solid urea and/or NH_4NO_3. All N solutions are classed either as pressure or non-pressure solutions.

  Pressure solutions have an appreciable vapor pressure of free NH_3. They may require specialized tanks and equipment, especially if their vapor pressure is high at operating temperatures. Pressure solutions must be applied below the soil surface to avoid NH_3 loss. Pressure solutions should not be applied in direct seed contact because of NH_3 damage to germination.

  Non-pressure solutions usually contain NH_4NO_3, urea, and water. These solutions can be handled without the use of high-pressure tanks and equipment. They contain

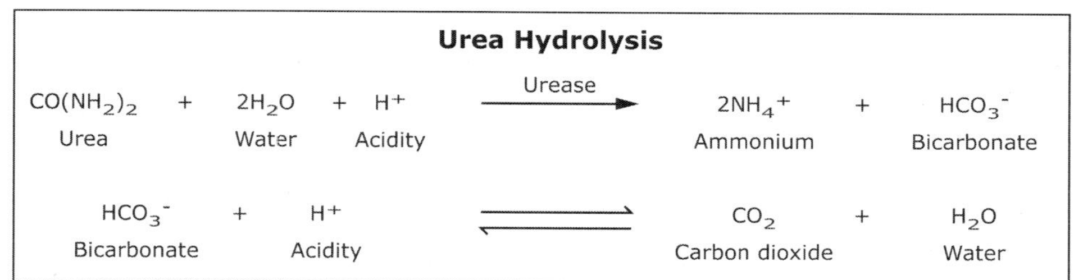

essentially no free NH_3. Nitrogen solutions containing both urea and NH_4NO_3, called urea ammonium nitrate (UAN), have higher N concentrations than solutions of either material alone. The presence of both compounds lowers the salt-out temperatures and allows the solutions to be used at lower temperatures without the formation of precipitates. Salting out...the precipitation of dissolved salts...occurs when the temperature drops to a critical point, the crystallization temperature. However, even in UAN solutions, as the N concentration increases, the salt-out temperature increases: 28% N salt-out is -5° F (-21° C); 30% N salt-out is 20° F (-7° C), and 32% N salt-out is 29° F (-2° C).

- **Ammonium nitrate** contains 33.5 to 34.0% N. Half the N in NH_4NO_3 is NH_4^+-N and half is NO_3^-–N. Although solid NH_4NO_3 has excellent handling qualities, it absorbs moisture (it is hygroscopic). For that reason, it is coated during manufacturing with materials such as diatomaceous earth to prevent absorption of water. Also, it should not be left in open sacks or bins for long periods in humid climates. Ammonium nitrate is well suited to bulk blending and to crops requiring sidedress N applications.

- **Urea** does not contain NH_4^+-N in the form in which it is marketed and used. However, in the soil it can quickly hydrolyze in the presence of the urease enzyme to produce NH_4^+ and bicarbonate (HCO_3^-) ions, **Figure 3-5**. A number of factors influence how quickly urea hydrolysis occurs, including the amount of the enzyme present and the soil temperature. The colder the soil, the slower the process **(Figure 3-6)**.

During hydrolysis, the HCO_3^- ions react with soil acidity and raise the soil pH in the vicinity of the urea hydrolysis reaction, offsetting some of the acidity produced later by nitrification. The NH_4^+ ions are adsorbed by soil clay and organic matter, eventually nitrified, or absorbed directly by plants. Once converted to NH_4^+-N, urea behaves like any other commercial N fertilizer and is an excellent source of N. There are some facts about urea behavior which should be understood, however.

Urea normally hydrolyzes rapidly. Significant quantities of NH_3 may be lost by volatilization when urea or urea-containing solutions are applied to bare soil surfaces which are rapidly evaporating water or to soils with large amounts of surface residues, including sod. Application during cold

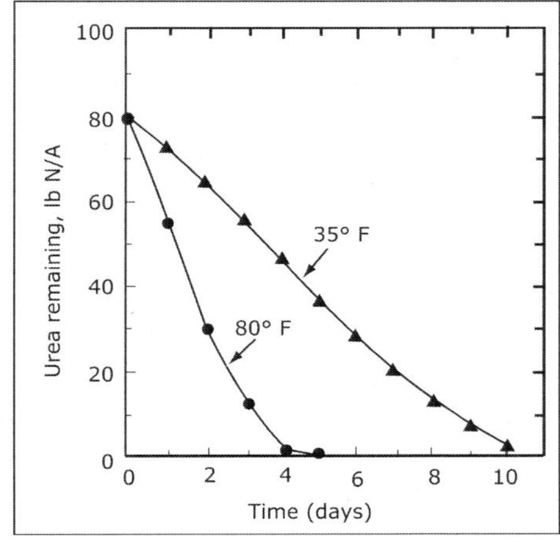

Figure 3-6. The conversion of urea to plant-available NH_4^+-N is dependent on soil temperature.

weather, soil incorporation, or surface band applications of urea-containing fertilizers (UAN) can help control the problem.

Rapid urea hydrolysis in soils may be responsible for NH_3 injury to seedlings when large quantities are placed too close to the seed or banded with the seed in narrow rows. A good rule of thumb is to avoid direct seed contact in row crops and narrow row-seeded small grains. Small grains can withstand higher rates of application in seed contact if the seeder lays down a wide band of seed and fertilizer (air seeders).

Urea is an excellent fertilizer for foliar application, but some grades may contain small amounts of a condensation product known as biuret. Biuret is toxic when applied to foliage, but has no detrimental effect when applied to the soil.

- **Ammonium sulfate** contains 21% N and 24% S. It is often produced as a by-product in the manufacturing of coke and nylon. Increasing frequency of S deficiencies is resulting in its wider use as a source of N and S.

- **Ammonium phosphates**—Monoammonium phosphate ($NH_4H_2PO_4$) and diammonium phosphate [$(NH_4)_2HPO_4$] are generally considered more important sources of P than N. These materials are discussed in Chapter 4.

Common N fertilizers, along with their N contents, are shown in **Table 3-9**. SFM

Table 3-9. Common N sources and their N contents.

Source	Percent N
Anhydrous ammonia	82
Aqua ammonia	20-25
Ammonium nitrate	33.5-34
Ammonium nitrate-sulfate	26
Ammonium nitrate/lime	20.5
Ammonium sulfate	21
Urea-ammonium nitrate solution (UAN)	28-32
Ammonium chloride	26
Urea	46
Monoammonium phosphate (MAP)	10-11
Diammonium phosphate (DAP)	18
Sodium nitrate	16
Potassium nitrate	13
Calcium nitrate	15.5
Sulfur-coated urea	39
Urea-formaldehyde	38

Phosphorus

Plants must have phosphorus (P) for normal growth and maturity. Phosphorus plays a role in photosynthesis, respiration, energy storage and transfer, cell division, cell enlargement, and several other processes in plants. Symptoms of deficiency may include stunted growth, reduced yield, and various other signs. Soil P comes largely from the weathering of apatite, a mineral containing P, calcium (Ca), and other elements. Phosphorus moves very little in most soils. Several factors affect P availability, including amount and type of clay, time of application, temperature, and other conditions. Placement of P fertilizer is an important consideration. Phosphate rock (PR) is the basic material used in all P fertilizer production. Understanding water solubility of various P fertilizers is important.

An Essential Plant Nutrient

Phosphorus is essential for plant growth. No other nutrient can be substituted for it. The plant must have P to complete its normal production cycle. It is one of the three primary nutrients. The other two are nitrogen (N) and potassium (K). **Table 4-1** shows some crops and the amounts of P (expressed as P_2O_5) they take up from the soil.

Table 4-1. Phosphorus uptake by some common crops.

Crop	Yield level	P_2O_5 taken up in total crop, lb	
Alfalfa	8	tons	120
Coastal bermudagrass	8	tons	96
Corn	160	bu	91
Cotton	1,000	lb lint	51
Grain sorghum	8,000	lb	84
Oranges	540	cwt	55
Peanuts	4,000	lb	39
Rice	7,000	lb	60
Soybeans	60	bu	58
Tomatoes	40	tons	87
Wheat	60	bu	41

Note: Phosphorus content of fertilizers is expressed as P_2O_5 equivalent, even though no P_2O_5 as such occurs in fertilizer materials. The P_2O_5 designation is a standard expression of relative P content. In this text, some results are reported in terms of P; others as P_2O_5. To convert P to P_2O_5, multiply by 2.29; to convert from P_2O_5 to P, multiply by 0.44.

<table>
<tr><td>Review Questions:
1. (T or F) Plants can complete their normal production cycle without P.</td></tr>
</table>

Roles of Phosphorus in Plants

Plants absorb most of their P as the primary orthophosphate ion ($H_2PO_4^-$). Smaller amounts of the secondary orthophosphate ion (HPO_4^{2-}) are taken up. Soil pH greatly influences the ratio of these two ions taken up by the plant. Other P forms may be utilized, but in much smaller quantities than the orthophosphates. Highest levels of P in young plants are found in tissue at the growing point. Since P moves readily from older to newer tissue, deficiencies first appear on the lower parts of plants (See **Appendix A**). Also, as crops mature, most P moves into the seeds (and/or fruits), **Table 4-2**.

Table 4-2. Seeds contain more P than other plant parts.

Crop	Part	Yield level	P content, %
Corn	Grain	150 bu	0.22
	Stover	7,500 lb	0.17
Cotton	Seed	2,000 lb	0.66
	Stalks	2,500 lb	0.24
Peanuts	Nuts	4,000 lb	0.20
	Vines	6,400 lb	0.26
Rice	Grain	6,000 lb	0.28
	Straw	7,000 lb	0.09
Soybeans	Grain	50 bu	0.42
	Straw	7,000 lb	0.18
Wheat	Grain	60 bu	0.42
	Straw	5,400 lb	0.12

Phosphorus is involved in photosynthesis, respiration, energy storage and transfer, cell division, cell enlargement, and several other processes in the plant. It promotes early root formation and growth. Phosphorus improves the quality of fruit, vegetable, and grain crops and is vital to seed formation. It is involved in the transfer of heredity traits from one generation to the next.

Phosphorus helps roots and seedlings develop more rapidly and improves winter hardiness. It increases water use efficiency, contributes to disease resistance in some plants, and hastens maturity...important to harvest and crop quality. See **Table 4-3**.

Table 4-3. Phosphorus fertilization boosts corn yield and lowers grain moisture at harvest.

P_2O_5 rate, lb/A	Yield, bu/A	Grain moisture, %
0	99	31.8
40	131	27.8
80	141	27.0
120	135	26.9
160	139	26.5

Low P Soil — Illinois

An important aspect of soil P fertility is its influence on P uptake by crops during periods of moisture stress. **Figure 4-1** shows that P uptake by corn seedlings is reduced during periods of moisture stress. However, the effect of the stress (drought) can at least be partially overcome when soil P levels are high.

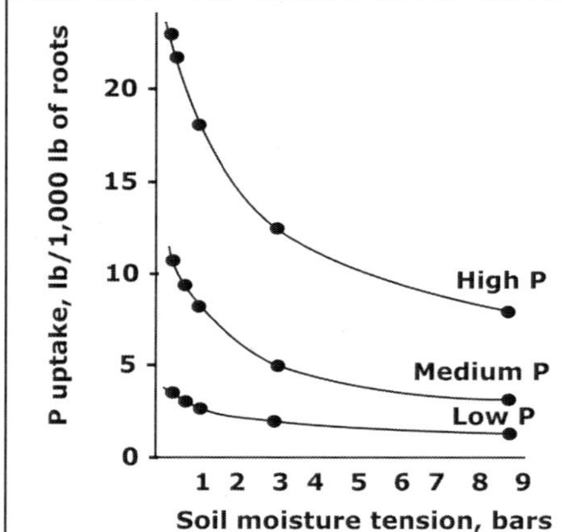

Figure 4-1. Soil P level affects uptake by corn plants during periods of moisture stress.

Review Questions:

2. The two most common forms of P taken up by plants are ______ and ______.

3. (T or F) Soil pH influences the ratio of P forms plants absorb.

4. (T or F) Phosphorus hastens crop maturity.

5. The plant parts usually highest in P are the ______ and/or ______.

Plant Deficiency Symptoms

The first sign of P hunger is an over-all stunted plant. Leaf shape may be distorted. With severe deficiency, dead areas may develop on the leaves, fruit, and stems. Older leaves will be affected before younger ones. A purple or reddish color associated with accumulation of sugars is often seen on deficient corn plants and on some other crops, especially during low temperatures (see **Appendix A**). Phosphorus deficiencies result in delayed maturity. Small grains grown on soils without adequate P tiller less.

Visual deficiency symptoms, other than stunted growth and reduced yield, are not as clear as are those for N and K. Phosphorus deficiency is difficult to detect in many field crops. At some growth stages, it may cause the crop to look darker green. One should always be alert for the characteristic stunted condition and, when possible, confirm what the eye sees with soil and/or plant analyses.

Phosphorus can be added to the soil P reservoir in inorganic and organic forms. Animal manures, biosolids, and plant residues are organic. Commercial fertilizers and primary soil minerals are inorganic. Phosphorus is released from minerals by natural weathering processes. It is also made available as soil organic matter is mineralized.

There are several ways P can be lost from the soil, including that removed by crop harvest. Because P is so strongly bound to the solid phase of the soil, it can be lost through soil erosion. Phosphorus in the soil solution is lost when runoff occurs, and P can also be lost by leaching under some conditions.

Figure 4.2 shows the soil P cycle. Phosphorus transformations shown in **Figure 4.2** are discusssed in the following sections of this chapter.

Figure 4-2. The soil P cycle.

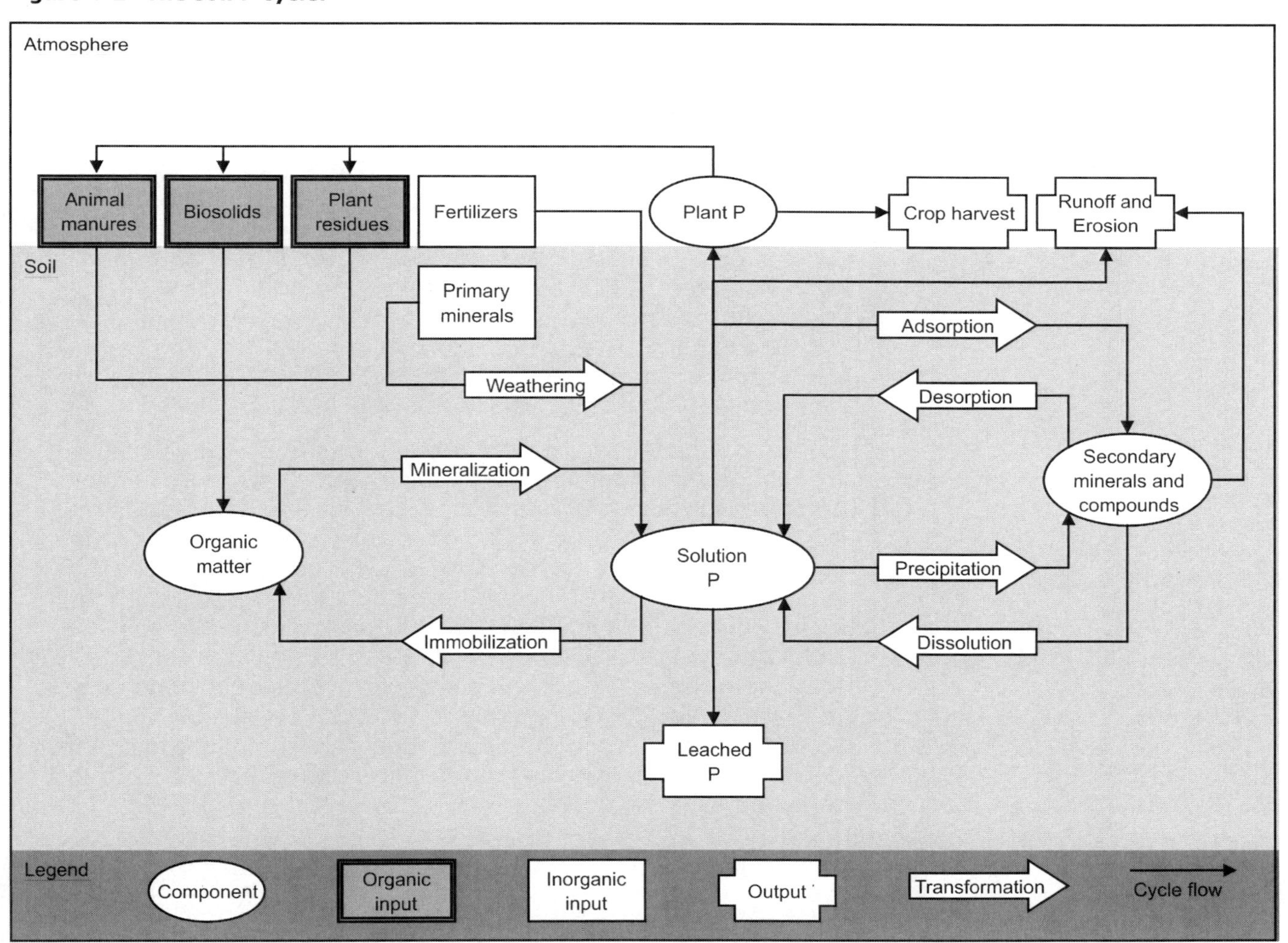

Elemental P is very reactive chemically, so it is not present in the pure state in nature. It is found only in chemical combinations with other elements. Soil P comes largely from the weathering of apatite, a mineral containing P and Ca, along with other elements such as fluorine (F) and chloride (Cl). As apatite breaks down and as organic matter decomposes and they release P into the soil, several P compounds are formed. The two orthophosphate ions taken up by plant roots are released. These orthophosphate ions are present in small amounts in the soil solution.

Phosphorus in soil solution reacts strongly with surfaces of secondary soil minerals and other compounds. Two primary reactions, adsorption and precipitation, are believed to occur. With adsorption, P is strongly held to mineral surfaces. In precipitation reactions, P can form insoluble compounds with other elements. Both types of reactions leave very little P in soil solution. However, a portion of adsorbed and precipitated P may re-solubilize through desorption and dissolution reactions.

Soluble P in the soil will form compounds with Ca, iron (Fe), and aluminum (Al), whether it comes from apatite, fertilizer, manure, or organic matter. Most of these compounds are not available to plants because they are insoluble. They are said to be in reverted or fixed forms. However, compounds such as dicalcium phosphate and octacalcium phosphate formed from the reaction of P fixed with soil Ca are relatively available to plants. Other plant available-sources of P include decomposing organic matter, humus, microorganisms, and other life forms. Research shows that organic compounds in the soil can help delay P fixation reactions.

The plow layer of most agricultural soils contains 800 to 1,600 lb of P/A in combination with other elements, most of it in forms not available to plants. Only a very small amount of total P in the soil is in solution at any one time...usually less than 4 lb/A. However, a few pounds of P/A in the soil solution are usually adequate for normal crop growth. The key to P fertility is not large quantities of P in the soil solution, but the soil's ability to replenish that solution P quickly.

As roots penetrate the soil mass and take up the available P, it must be replaced on a continuous basis. Phosphorus in soil solution is replaced about twice a day, on average, but as many as 10 times a day...about 300 times or more during the growing season of crops such as corn and soybeans. A soil must replenish or maintain sufficient levels of solution P to insure high yields. **Figure 4-3** shows: 1) how P is replenished in soil solution, 2) how it is made unavailable and 3) how it is removed (or lost) from the soil. Note the double arrow between "Phosphorus in Soil Solution" and "Minerals." REMEMBER: Phosphorus becomes available through mineral weathering and organic matter decomposition, but it can also become unavailable or fixed in a form the plant can't use.

Phosphorus

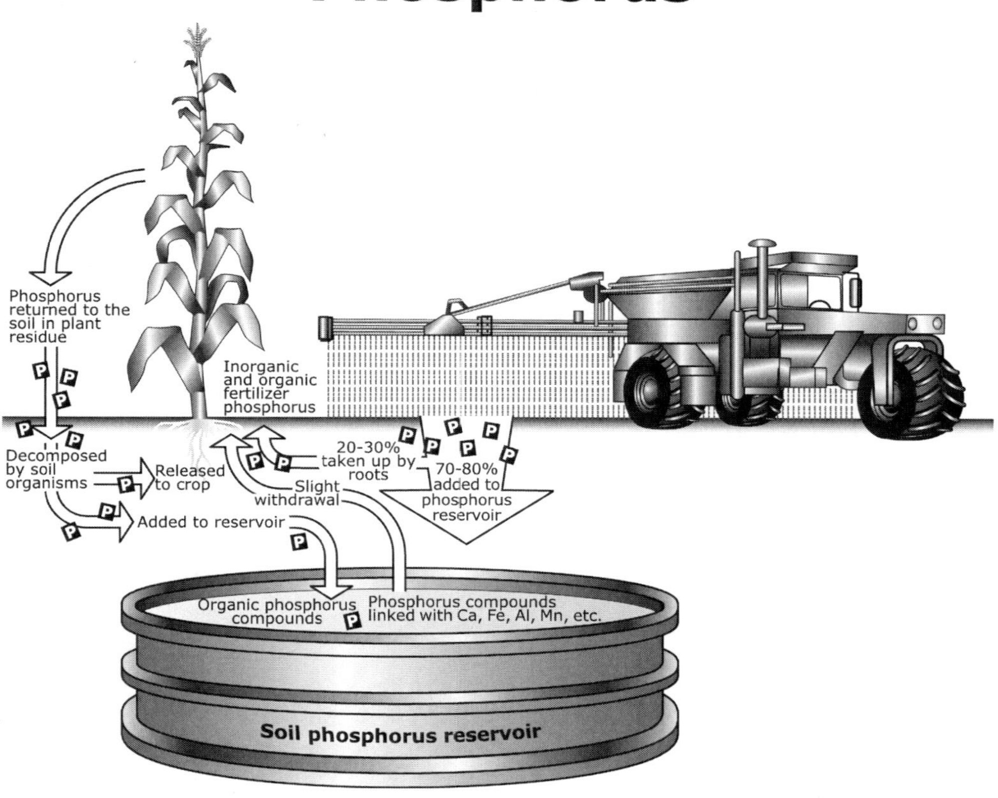

Most soils don't have enough...

Most crops have a difficult time getting enough P. Phosphorus deficiency may be more limiting in world crop production than other deficiencies, toxicities, and diseases. A 2005 Potash & Phosphate Institute summary of soil test information indicated that the median P level for North America is 31 ppm, with 42% of samples testing less than 25 ppm, a "middle-of-the-road" critical level. Here are some samples of median Bray P-1 equivalent soil test levels.

A grower can expect to get 10 to 30% efficiency from mineral P fertilizer the first year after application. It is difficult to keep P available to plants. It easily combines with such elements as Ca and Fe to form compounds that are less available to plants. Methods of application can influence P use efficiency.

State or province	Median Bray P-1 equivalent soil test levels
Iowa	25
Arkansas	17
Pennsylvania	57
Texas	18
Nebraska	22
Saskatchewan	10
Minnesota	18
Quebec	37
Indiana	29
California	24

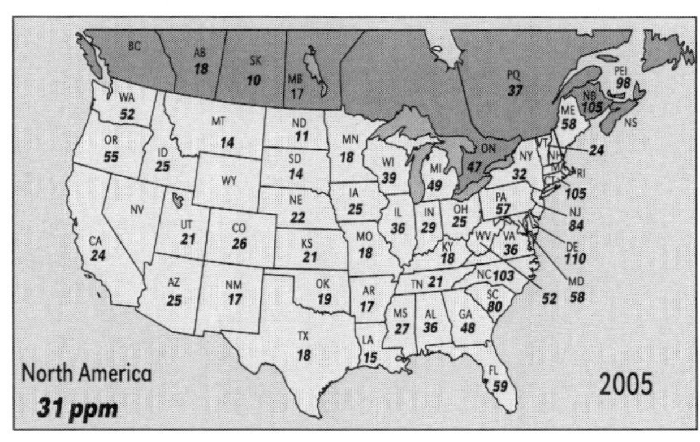

Median Bray P-1 equivalent soil test levels.

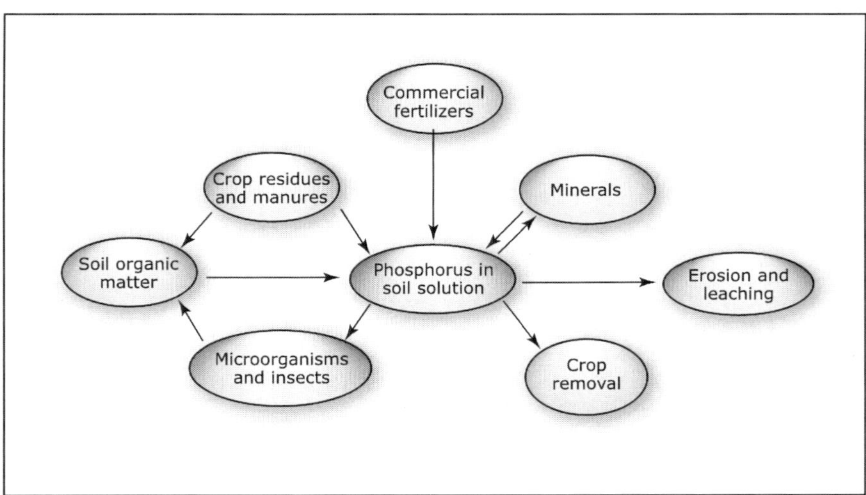

Figure 4-3. Phosphorus content of the soil solution is affected by many factors.

Phosphorus Movement in Soils

Phosphorus moves very little in most soils. It generally stays where it is placed by mineral weathering or by fertilization. Little P is lost by leaching, though it moves more freely in sandy than in clay soils. Surface erosion (run-off) can remove soil particles containing P. Erosion and crop removal are the only significant ways soil loses P.

Nearly all P moves in the soil by diffusion, a short-ranged process that depends on soil moisture and temperature. Dry conditions and soil temperatures reduce diffusion sharply. Potassium also moves largely by diffusion, but it is more soluble than P so it tends to travel further. When the distances N, P, and K move from the point of placement are compared, it is apparent how freely N as nitrate (NO_3^-) moves in the soil. This comparison is only relative, not absolute (**Figure 4-4**).

How little does P actually move? If the P in a loamy soil is more than ¼ in. from a root, it will not move close enough to be taken up by the root. It has been estimated that roots of a growing crop contact 3% or less of the soil in the surface 6 to 7 in. In practical terms, this means soil must be adequately supplied with P to support optimum crop growth. The soil P level throughout the root zone should be high enough to insure available P during every stage of growth.

The importance of season-long P availability cannot be overemphasized. Data in **Table 4-4** show the per-day and season-long uptake of P_2O_5 by soybeans which yielded 100 bu/A. During the first half of the growing season (51 of 103 days), only 9% of total P_2O_5 was taken up. That means 91% or 120 lb/A were taken up in the last 52 days.

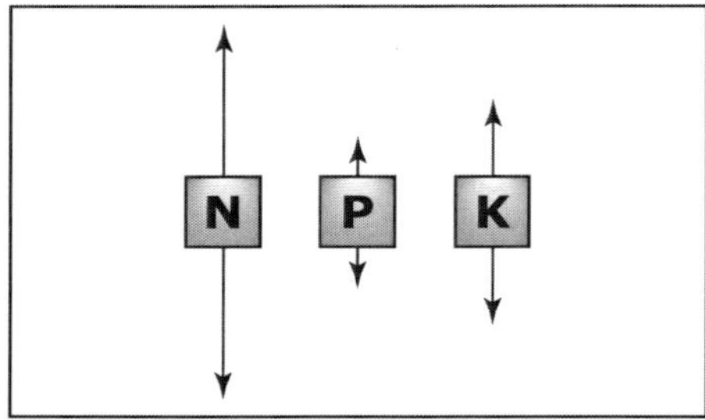

Figure 4-4. Relative movement of N, P, and K in the soil.

Table 4-4. Soybeans take up P_2O_5 throughout the growing season.

Growth stage	Days	P_2O_5 uptake, lb/A		Percent of total uptake
		Per day	Total	
Emergence to 3-leaf	40	0.15	6	4.5
3-leaf to 6-leaf	11	0.55	6	4.5
6-leaf to full bloom	16	1.75	28	21.2
Full bloom and early pod development	15	2.27	34	25.8
Pod filling to mature seed	21	2.76	58	43.9
Totals	103	—	132	100.0

New Jersey

Review Questions:

11. (T or F) Phosphorus moves freely in the soils.

12. Soil P is (more, less) mobile than soil K.

Factors Affecting Phosphorus Availability

Most crops recover only 10 to 30% of mineral fertilizer P during the first year following application (See **Production Concept 4-1**). Recovery varies widely, depending on P source, soil type, crop grown, application method, and weather. But much of the residual P will be available to succeeding crops. Phosphorus availability varies with the following factors:

1. **Amount of clay**–Soils high in clay content will fix more P than those containing less clay.

2. **Type of clay**–Soils high in kaolinitic clays (common to high rainfall and temperate regions such as the southern U.S.) fix more added P than other soils. Regardless of clay type, fertilizer P is rapidly converted to less available forms.

3. **Time of application**–The longer the soil and added P are in contact, the greater the chances for fixation. On high-fixing soils, the crop must use fertilizer P before fixation sets in. On other soils, P utilization may occur for years. This critical period...how long after application the plant can effectively utilize fertilizer P...determines the P fertilization schedule. Should P be applied occasionally in larger amounts, as in a rotation? Or should it be applied more frequently in smaller amounts? Time and method(s) of P fertilization should be matched to the cropping system to insure most efficient use.

4. **Method of application** – See the following section for discussion.

5. **Aeration**–Oxygen (O_2) is necessary for plant growth and nutrient absorption. It is also essential for microbiological breakdown of soil organic matter, an important P source.

6. **Compaction**–Compaction reduces aeration and pore

space in the root zone. This reduces P uptake and plant growth. Compaction also decreases the soil volume plant roots penetrate, limiting their total access to soil P. The fact that P moves such short distances in most soils adds to the problem of restricted root growth and nutrient uptake brought on by compaction.

7. **Moisture**–Increasing soil moisture to optimum levels makes P more available to plants. But excess moisture reduces O_2, limiting root growth and slowing P uptake.

8. **Phosphate status of soil**–Soils that have received more P fertilizer than crops have removed for several years may show an increased level of available P, enough to reduce current fertilization if the soil level is high. In fact, under some combinations of high soil P and environmental conditions, P fertilization might need to be suspended. It is important, however, to monitor soil P and maintain optimum levels to support optimum crop production.

9. **Temperature**–When temperatures are right for good plant growth, they affect P availability very little. High temperatures encourage organic matter decomposition. But when temperatures are too high or too low, they can restrict P uptake by the plant. That is why crops often respond to starter P on cold, wet soils, even when soil P levels are optimum. Fertilizer P solubility is not affected by low soil temperature.

10. **Other nutrients**–Applying other nutrients may stimulate P uptake. Calcium on acid soils and sulfur (S) on alkaline soils seem to increase P availability, as does ammonium-N (NH_4^+-N). But zinc (Zn) fertilization with border-line P deficiency tends to restrict P uptake further. Potassium applications have little or no effect. (See **Production Concept 4-2**.)

11. **Soil pH**–The solubility of various P compounds in the soil is largely determined by pH. Phosphates of Fe, Al, and manganese (Mn) have low water solubilities. They dominate acid soils. Insoluble Ca and magnesium (Mg) compounds exist above pH 7.0. The more soluble or available P forms exist in the 5.5 to 7.5 range. This makes sound liming essential on very acid soils. (See **Production Concept 4-3**.)

12. **Crop**–Some crops have fibrous root systems; others are tap rooted. Wheat has a relatively shallow root system while alfalfa explores deep into the soil profile. Therefore, crops differ greatly in their ability to extract available P from the soil.

Nitrogen Improves Phosphorus Uptake

EARLY PLANT GROWTH should be strong and fast to get the plant well established before summer rigors set in...dry periods, insects, weeds, etc.

Phosphorus is vital to early growth. And N influences plant uptake of P.

When applied with N, P is more available to plants than when applied without N.

This influence of N on P uptake is very clear during early growth. In some cases, as much as 65% of P in the plant comes from fertilizer early in the season.

Ammonium N has significant effects on P availability and absorption. In high concentrations, NH_4^+–N slows P fixation reactions. Ammonium absorption helps maintain an acidic condition at the root surface, improving P absorption.

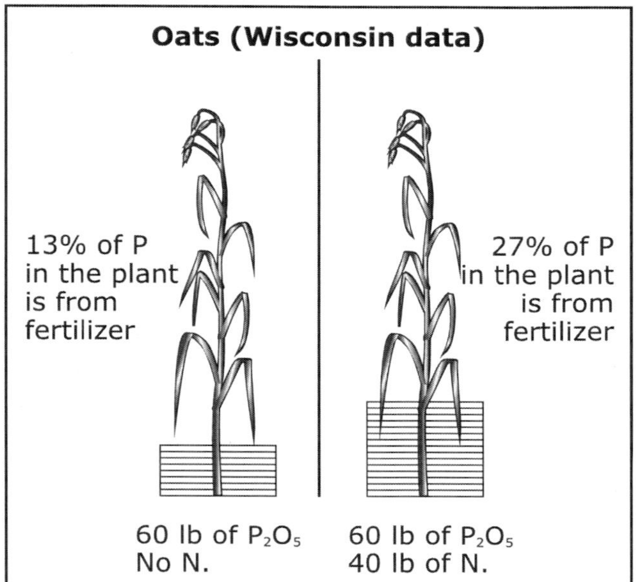

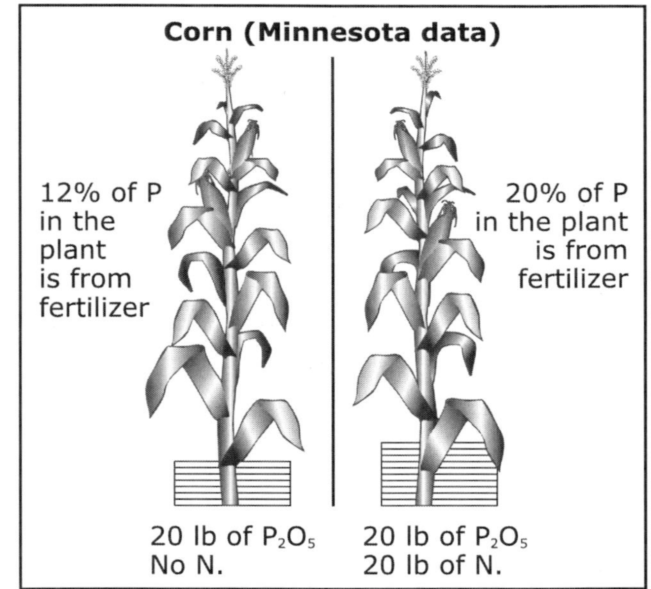

Availability of Phosphorus Varies with Soil pH

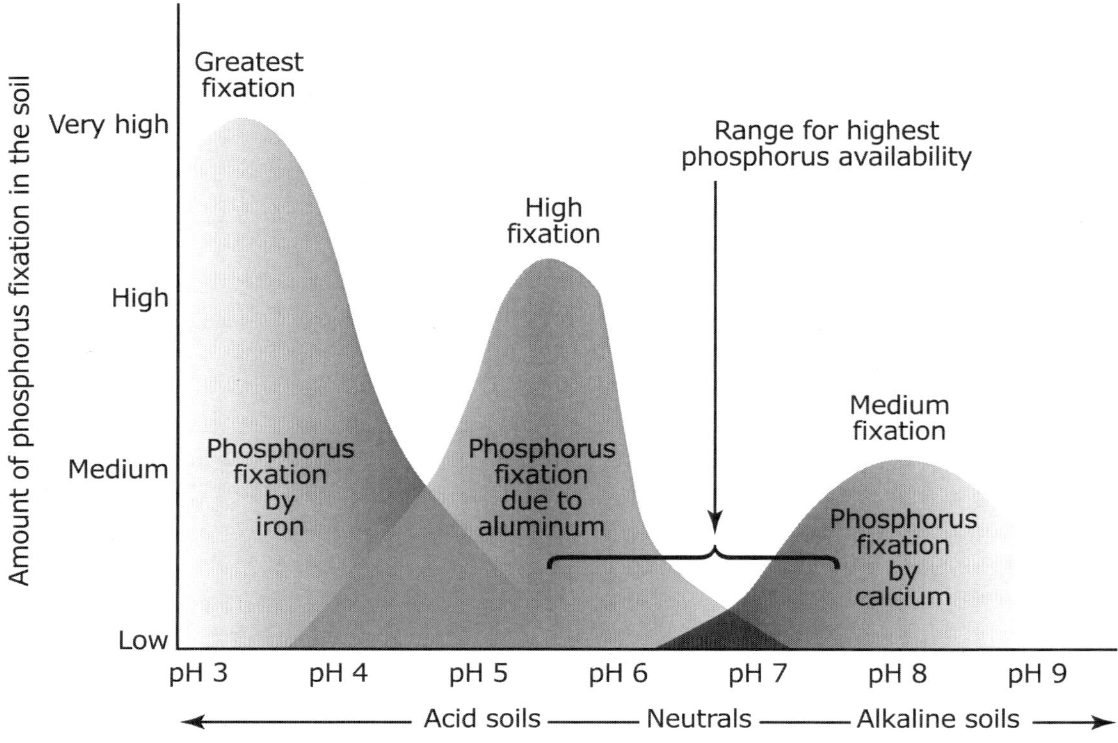

Soil pH greatly influences the solubility of different P compounds in the soil. Solubility indicates how available the P is, or how fixed or tied up it becomes in the soil. In acid soils (decreasing pH), P reacts with Fe, Al, and Mn to form insoluble products, making P less available. In alkaline soils (increasing pH), Ca and Mg react with P to lessen P availability as pH increases above 7.5. The more soluble or available forms exist in the range of pH 5.5 to 7.5. This makes liming essential on very acid soils. Lowering the pH of alkaline soils specifically to improve P availability is not considered practical.

**Phosphorus is
most available
from pH 5.5 to pH 7.5**

There is no set method of applying P fertilizer. Many factors must be considered, including soil fertility levels, crop(s) to be grown, tillage methods, equipment, timing, and other management factors. Phosphorus fixation is an important point to consider when deciding how to apply P. More P fertilizer contacts more soil when it is broadcast and plowed down or disked in than when it is banded, so the amount of fixation is higher.

Crops usually respond more to banded than to broadcast P on low fertility soils. Fixation is greater when fertilizer is broadcast. Further, banding puts a readily accessible P source in the root zone, making it positionally more available. Banding also concentrates other nutrients with the P such as NH_4^+-N, which can slow fixation reactions and enhance P uptake.

If a grower is looking for maximum return from P investment, band application is the best bet. But, as fertility levels increase, the banding advantage disappears, and potential yields go up. So whether to band or broadcast depends greatly on the management philosophy of the grower. Does the grower fertilize for only maximum returns in the short run or instead build long-term opportunity for higher yields and profits by raising soil test P levels? Land tenure has much to do with that decision.

Broadcast and/or plow down applications have several advantages:

- High rates can be applied without injuring the plant.

- Nutrient distribution throughout the root zone encourages deeper rooting, while band placement causes root concentration around the band.

- Deeper rooting permits more root-soil contact, providing a larger reservoir of moisture and nutrients.

- The most practical way to topdress P on established pastures and meadows.

- Insures high level fertility to help the crop take full advantage of favorable conditions throughout the growing season.

- Can be done at times other than during the busy planting season.

Banding is time consuming when applied through the planter or drill (starter), and it is difficult to apply large amounts of fertilizer using this method. However, preplant banding, particularly of N, P, and S, has been very effective. Banding does offer several advantages:

- May allow use of lower rates than broadcast to achieve the same yield levels.

- Is advantageous to a renter who might have only a short-term lease and does not wish to build soil fertility levels which will cost the renter extra, but might benefit others.

- Reduces fixation of P.

- Places the P so that it is positionally available to the young, restricted root system.

NPK Placement and Movement

Nitrogen
Movement in surface soil profile

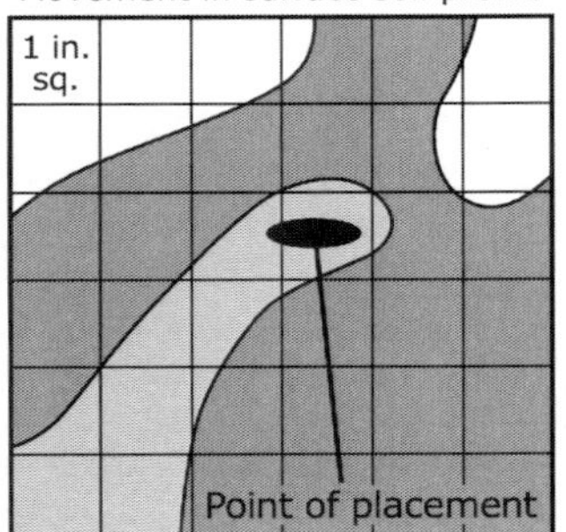

Nitrogen location 17 days
after application

Nitrogen moves through the soil rather freely during the growing season. Positioning N in the root zone is generally not critical for root interception in conventional tillage systems. However, band placement of N has been shown to significantly enhance N use efficiency under reduced tillage conditions. Band placement of N can also slow the nitrification process.

Phosphorus
Movement in surface soil profile

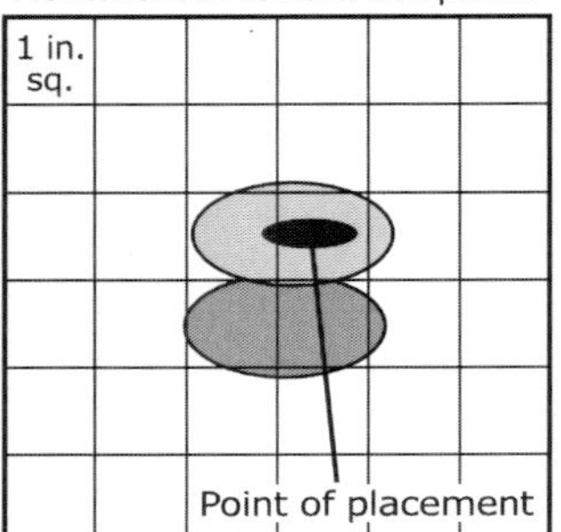

Phosphorus location 17 days
after application

Phosphorus needs most attention with its placement. This illustration shows how very limited its movement is. Phosphorus should be placed where roots can intercept it. Banding P is the most agronomically efficient way to place it in low fertility soils. Banding $NH_4^+–N$ with P enhances P uptake.

Potassium
Movement in surface soil profile

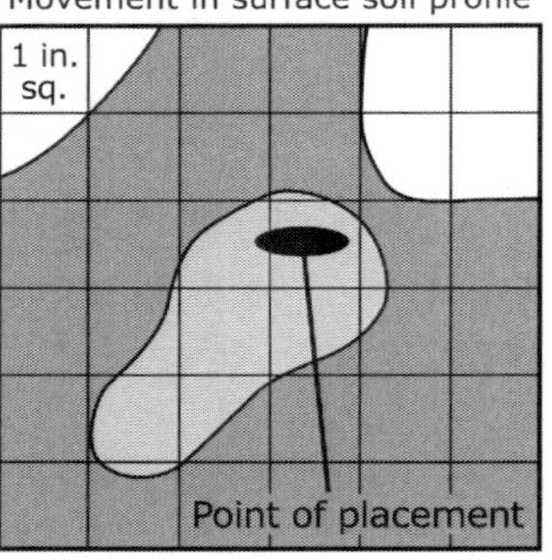

Potassium location 17 days
after application

Potassium placement is critical. Like P it does not move readily in the soil. Broadcast application is generally most effective, sometimes in combination with band placement. Band placement of K in conservation tillage systems can significantly improve K availability, probably related to plant rooting patterns. Deep banding K has been important in helping overcome subsoil K deficiency in cotton.

High concentration

Medium concentration

No effect

NPK movement in surface soil adapted from Michigan State University data.

Table 4-5. Phosphorus placement can have significant effects on corn yields and P use efficiency.

Application method	Kansas corn, bu/A	Kansas sorghum, lb/A	Kansas wheat, bu/A	Nebraska wheat, bu/A	Alberta barley, bu/A	N. Dakota barley, bu/A
Broadcast	170	4,760	53	34	44	58
Preplant band	184	5,600	64	42	58	68

Low soil test P

- Provides an opportunity for enhanced use efficiency of P and at the same time increased yields by combining placement and recommended P rates.

Although surface-applied P is usually the least efficient way to fertilize row crops, no-till planting in relatively warm, humid areas is an exception. When a crop such as corn is planted in dead sod or in crop residues with no previous tillage, surface-applied P can be as effective as placed P. With surface residues, moisture levels encourage shallow rooting. This enables the roots to utilize P at or near the surface. Reduced tillage under conditions of low fertility, limited moisture, and cold soils places additional emphasis on P placement.

On low P soils and in colder areas, banded P is important for many crops in both conventional and reduced tillage systems. Research has shown that preplant banding, which produces zones of high P concentration, can significantly affect a plant's ability to use fertilizer P, improving yield and P use efficiency as shown in **Table 4-5**. High concentrations of P along with concentrations of NH_4^+ may slow P fixation reactions, boosting availability. Placement deeper into the soil can also benefit P uptake under dryland conditions. Combinations of preplant banding and starter P may be even more effective, especially when soils are cold.

Drilling P with small grain requires less fertilizer than broadcast to produce a given yield increase. But it can consume valuable time at critical periods. Placement directly under the drill row (band seeding) for forage crops surpasses broadcast or placement to the side of the row. Tomatoes and onions have responded best to P placed directly below the seed or set. Preplant dual application of anhydrous ammonia (NH_3) or urea-ammonium nitrate (UAN) solutions and ammonium polyphosphate (APP) for wheat has been found to be superior to broadcast P applications, particularly on lower P testing soils.

Sometimes row and broadcast applications are combined for best effect. This ensures an early, accessible P supply for developing seedlings and a nutrient reserve throughout the growing season. The starter effect from banding, even on high P soils, is often important when temperatures are low, a common condition for early-planted or reduced tillage crops. Some vegetable crops...potatoes, for example...respond to banded P even on very high P soils.

Review Questions:
17. Fixation is (higher, lower) when P is broadcast and plowed down compared to banding.
18. Drilling fertilizer P with small grain seed requires (more, less) fertilizer to produce a given yield increase than if it is broadcast.
19. The starter effect from banding, even on soils high in P, can be important when temperatures are ______.

Phosphate rock (PR) is the basic material used in all P fertilizer production. The most important deposits are sedimentary materials, laid down in beds under the ocean and later lifted up into land masses. About 25% to 40% of the reserve can be economically recovered under today's conditions with a reserve base at about 50 billion tons. This tonnage represents enough P to meet present consumption for many decades, or even hundreds of years. As the economy changes, even greater amounts can be recovered.

U.S. deposits (phosphorite) are found in Florida, North Carolina, Tennessee, Idaho, Montana, Utah, and Wyoming. They represent about 8% of the world's known reserves and 8% of the reserve base. Production from Florida represents nearly three-fourths of the U.S. total, the remainder coming from the western states, Tennessee, and North Carolina.

Almost all PR is strip mined. It usually contains about 34% P_2O_5 and must be upgraded for use as fertilizer. Upgrading removes clay and other impurities. This process is called beneficiation. Following beneficiation, the PR is finely ground. Although some is applied directly as fertilizers on acid soils in some countries, the P in it is slowly available, and large quantities must be applied (1 t/A or more) before much response can be expected. Usually, PR is treated to make the P a more soluble and reliable source.

Fertilizer phosphates are classed as either acid-treated or thermal-processed. Acid-treated P is by far the most important. Sulfuric (H_2SO_4) and phosphoric (H_3PO_4) acids are used in producing acid-treated phosphate fertilizers. Sulfuric acid is produced from elemental S or from sulfur dioxide (SO_2). More than 60% of this industrial acid is used to produce fertilizers. Treating PR with concentrated (about 90 to 93%) H_2SO_4 produces a mixture of H_3PO_4 and gypsum ($CaSO_4$). Filtration removes the $CaSO_4$ to leave green, wet-process, or merchant grade H_3PO_4 containing about 54% P_2O_5.

Wet process acid can be further concentrated to form superphosphoric acid (SPA). Superphosphoric acid is made by evaporation of water from wet process H_3PO_4. In this process, two or more orthophosphate molecules combine to form polyphosphate compounds. These polyphosphate products are commonly used for the manufacture of clear fluid fertilizers. They contain 68 to 80% P_2O_5.

Wet process acidulation is the most commonly used technique for solubilizing P in PR. The majority of P fertilizers used in North America are based on wet process H_3PO_4 (52 to 55% P_2O_5) resulting from the reaction of PR with H_2SO_4. Common P-containing fertilizers and their manufacturing processes are listed below.

- **Normal or single superphosphate (NSP or SSP)** is made by treating PR with a measured amount of about 60 to 72% H_2SO_4. Normal superphosphate contains about 20% P_2O_5 and 12% S. It no longer has widespread use, although it is a good source of both P and S. Because it absorbs NH_3, it has been used to produce ammoniated superphosphates.

- **Concentrated superphosphate (CSP)** or **triple superphosphate (TSP)** comes from reacting wet-process H_3PO_4 with PR. It contains approximately 46% P_2O_5.

- **Ammonium orthophosphates (AOP)** are produced by ammoniating H_3PO_4. Monoammonium phosphate (MAP: 10 to 12% N and 48 to 55% P_2O_5) and diammonium phosphate (DAP: 18-46-0) are made by controlling the amount of NH_3 reacted with H_3PO_4.

- **Ammonium polyphosphates (APP)** are usually fluid sources of P produced by ammoniating superphosphoric acid. Polyphosphate P_2O_5 content ranges from 40 to 70%. Common analyses of polyphosphate liquid fertilizers are 10-34-0 and 11-37-0.

- **Nitric phosphates (NP)** are made by acidulating PR with nitric acid (HNO_3). To make the material more water soluble, some H_2SO_4 or H_3PO_4 is used with the HNO_3. Most nitric phosphates are produced and used in European countries.

- **Ammoniated superphosphates** are made by reacting either NSP or TSP with NH_3. They are available in different fertilizer grades and water solubilities. The water soluble P in such fertilizers is influenced by phosphate source, degree of ammoniation, content of impurities (other salts), moisture content, speed of drying, etc.

Thermal phosphoric acid manufacture begins with the production of elemental P through the reduction of PR with coke in an electric arc furnace. Elemental P is oxidized to P_2O_5 which is subsequently reacted with water to form furnace grade H_3PO_4. Thermal acid is much purer than wet-process H_3PO_4. Its use in fertilizer manufacture is sometimes preferred for the production of liquid fertilizers because of its purity. Agronomically, furnace grade H_3PO_4 derived products and those produced from merchant grade H_3PO_4 are identical, including the reactions that these products undergo in the soil.

Phosphate Fertilizer Terminology

The water solubility of the P in a specific P fertilizer does not always tell how available it is. Chemical methods can rapidly determine the total P content, how much is water soluble, and how much is plant-available. Phosphate solubility in fertilizers is described as: water soluble, citrate soluble, citrate insoluble, available, and total.

- **Water soluble P** can be extracted from the fertilizer material using only water.

- **Citrate soluble P** can be extracted with a solution of 1 normal neutral ammonium citrate after water soluble P has been removed.

- **Available P** is the sum of the water soluble and citrate soluble fractions.

- **Citrate insoluble P** is the portion remaining after extraction with both water and ammonium citrate.

- **Total P** is the sum of available P and citrate insoluble P

Repeated research has demonstrated that as long as a P fertilizer contains 60% or more water soluble P, agronomic performance is essentially equal to fertilizers containing 100% water soluble P.

<table>
<tr><td>Review Question:
30. Available P is the sum of ______ ______ and ______ ______ fractions.</td></tr>
</table>

Summary

Research has shown that all common P fertilizer sources are similar agronomically when equal rates are applied and method of application is comparable. There are advantages/disadvantages, including handling and storage. Proper application to insure best availability and to prevent potential damage to seeds and seedlings should be followed. SFM

Chapter 5

Potassium

Potassium (K) is an essential nutrient and is taken up in significant amounts by many crops. Potassium is vital to photosynthesis, protein synthesis, and many other functions in plants. When plants are deficient in K, they may grow slowly, symptoms may appear on leaves, stalks or stems may be weak, and disease resistance may be lowered. In the soil, K exists in three forms: unavailable, slowly available, and available. Unlike some other nutrients, K moves very little in the soil. The K in fertilizer takes on the ionic form (K^+) when it dissolves. Many soil factors can affect K uptake by plants. Methods of applying K fertilizers range from banding to broadcast. Many different types and analyses of K fertilizers are available, originating from a wide range of sources.

An Essential Plant Nutrient

Potassium is a vital plant nutrient. No other nutrient can replace it. It is one of the three primary nutrients, along with nitrogen (N) and phosphorus (P). Agronomic crops contain about the same amount of K as N, but more K than P. In many high-yielding crops, K content exceeds N content. **Table 5-1** shows the amount of K_2O some crops take up from the soil.

Table 5-1. Potassium uptake by some common crops.

Crop	Yield level		K_2O taken up in total crop, lb
Alfalfa	8	tons	480
Coastal bermudagrass	8	tons	400
Corn	160	bu	213
Cotton	1,000	lb lint	85
Grain sorghum	800	lb	240
Oranges	540	cwt	330
Peanuts	4,000	lb	185
Rice	7,000	lb	168
Soybeans	60	bu	205
Tomatoes	40	tons	460
Wheat	60	bu	122

Note: Potassium content of fertilizers is expressed as K_2O equivalent, or potash, even though no K_2O as such occurs in fertilizer materials. The K_2O designation is a standard expression of relative K content. In this text, some results are reported in terms of K, others as K_2O. To convert from K to K_2O, multiply by 1.2. To convert from K_2O to K, multiply by 0.83.

Review Questions:

1. Potassium is one of three _______ plant nutrients. The other two are _______ and _______.

2. Most crops contain (more, less, about the same) K compared to P.

3. To convert pounds of K to pounds of K_2O, multiply by _______: to convert K_2O to K, multiply by _______.

Potassium Plays Many Roles in the Plant

Potassium is absorbed (taken up from the soil) by plants in the ionic form, indicated as K^+. Although it is essential for plant growth, many of its functions in the plant are not well understood. Unlike N and P, K does not form organic compounds in the plant. Its primary function is related to ionic strength of solutions inside plant cells. It provides much of the osmotic pull that draws water into plants.

Potassium is vital to photosynthesis. When K is deficient, photosynthesis declines, and the plant's respiration increases. These two K-deficient conditions—reduced photosynthesis and

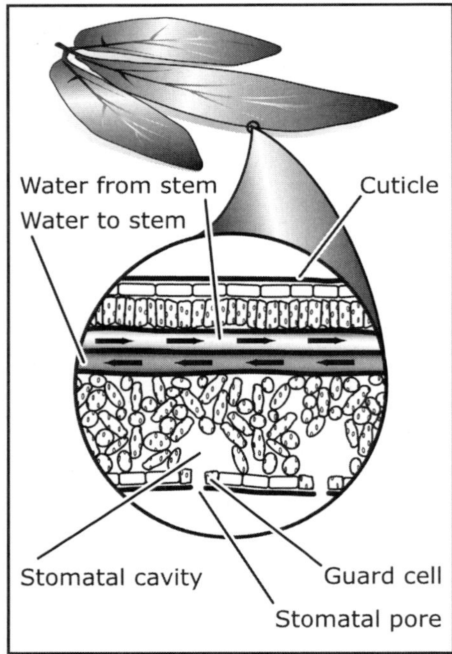

Figure 5-1. Opening and closing of stomatal pores in plant leaves is regulated by K concentration in guard cells surrounding the stomates.

increased respiration—lower the plant's carbohydrate supply. Other functions of K...

- essential for protein synthesis.
- important in the breakdown of carbohydrates, a process which provides energy for plant growth.
- helps to control ionic balance.
- helps the plant to overcome the effects of diseases.
- improves winter hardiness.
- activates more than 80 enzyme systems that regulate the rates of major plant growth reactions
- provides turgor; K-deficient plants wilt more easily.

An important role of K in plant growth is its influence on water use efficiency. The process of opening and closing of plant leaf pores, called stomates, is regulated by K concentration in the guard cells which surround the stomates (**Figure 5-1**).

When stomates open, large quantities of K move from the surrounding cells into the guard cells. As K moves out of the guard cells into surrounding cells, stomates close. Thus, K plays a key role in the process plants use to conserve water.

A shortage of K causes the stomates to open only partially and to be slower in closing. This increases the stress from drought. **Table 5-2** shows how adequate K improves yields of corn under three levels of rainfall.

Table 5-2. Potassium increases yields of corn and protects against abnormal moisture conditions.

Rainfall level		Rainfall during growing season, inches	Yield, bu/A		Increase, bu/A
			Low K	High K	
Low	(Indiana)	7.1	91	130	39
Medium	(Indiana)	17.7	148	156	8
High	(Indiana)	25.7	92	140	48
Low	(Ohio)	9.0	122	164	42
Medium	(Ohio)	19.9	151	173	22

The importance of K in disease suppression and crop quality cannot be over stated. See **Production Concept 5-1.**

Data in **Table 5-3** show how K increased yields and controlled lodging in corn. Lodging, frequently associated with stalk rot, was reduced in each of the four years of the study.

Table 5-3. Potassium increases corn yields and reduces lodging.

Year	Yield, bu/A		Stalk lodging, %	
	No K	120 lb/A K_2O	No K	120 lb/A K_2O
1	148	164	56	50
2	148	164	30	25
3	151	187	30	16
4	104	120	52	27

High K Soil   Illinois

POTASSIUM
The Quality Nutrient

Potassium (K) is involved in regulating many processes in plants. It is a major element in determining crop yield as well as quality. For many years, scientists have recognized its role in strengthening the natural disease-resisting mechanisms of plants.

Here are some examples from research studies, to illustrate the many benefits of K.

- **Potassium strengthens** plant stalks and stems against invading organisms and lodging. In cereal crops, it thickens the cuticle against mildew and other infections, making plant cells more turgid and less suitable for certain diseases to invade. Potassium is also credited with reducing verticillium wilt in cotton, mold and mildew in soybeans, leafspot and dollarspot in grasses, and numerous other conditions.

- **Protein synthesis requires K**. Without adequate K, raw nitrogen (N) materials such as amino acids and nitrate accumulate in the plant. Regardless of how much N is applied, K-deficient plants cannot convert it into protein. This leads to low quality grain and forage.

- **Potassium plays a role in activating at least 80 enzyme systems within the plant**. Inadequate K level disrupts development and plant functions in numerous ways.

- **Carbohydrate synthesis and transfer needs K**. If photosynthesis is slowed by lack of K, production of carbohydrates suffers, directly affecting the quality of crops such as potatoes, sugarbeets, grapes, and stone fruits.

- **Starch synthesis requires K**. Without adequate K, soluble carbohydrates are not converted to starch. Perennial crops, such as alfalfa, are more subject to winter-kill without sufficient root reserves of starch.

- **Potassium improves the shelf-life of bananas and many other fruits and vegetables.** It also helps the fiber quality of cotton, including the strength, length, and uniformity.

- **Potassium nutrition also shows promise related to beneficial nutraceutical (phytochemical) content of some crops.** Recent studies found that when soybean yields respond to K, isoflavone levels also increase. Lycopene content of tomatoes is also enhanced by K.

Potassium has a great impact on crop quality, including increased kernel weight and kernels per ear in corn, improved oil content in soybeans, better milling and baking quality in wheat, and improved stand and longevity in forages.

One of the problems in forage production is poor fertilizer management, particularly with regard to N and K fertilizer balance. Growers use N because they know it increases yield, adds green color, and improves protein content. Potassium is less showy than N and is often neglected. **Table 5-4** shows why balancing N and P with K is so important. **Figure 5-2** shows a similar relationship between N and K for timothy forage yields.

Table 5-4. Potassium increases yield and reduces leaf spot infection on Coastal bermudagrass.

Nutrients applied, lb/A $N-P_2O_5-K_2O$	Disease rating	Dry forage yield, lb/A (second cutting)
500-0-0	3.8	2,693
500-70-0	3.9	2,887
500-0-60	1.4	4,509
500-0-120	1.0	4,679
500-70-60	1.5	4,267
500-140-140	1.1	4,999

A rating of 1.0 is disease-free.   Texas

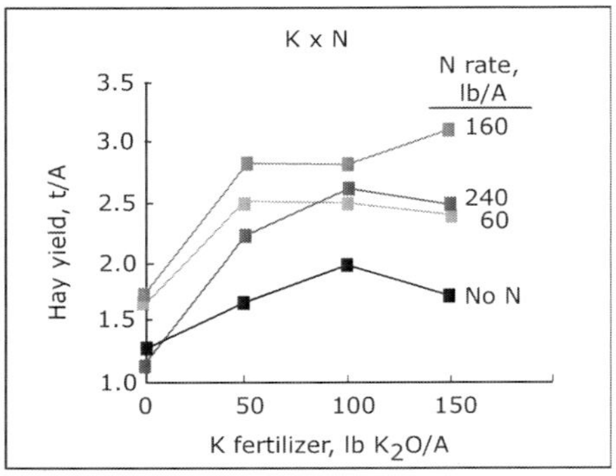

A large part of the reason K increases forage yields and controls diseases is because it improves the crop's winter hardiness. It allows crops to get a quicker start in the spring and increases vigor so growth can continue throughout the growing season. **Table 5-5** documents that K increases winter survival and provides season-long stand advantage of Coastal bermudagrass.

Table 5-5. Potassium increases stand of Coastal bermudagrass.

K_2O rate, lb/A	Percent stand	
	Spring	Late summer
0	43	17
120	73	59
240	69	88

    Texas

As plant roots explore the soil profile, they can run into many unfavorable conditions or stresses...moisture stress, physical and chemical barriers, insects, diseases. All these factors reduce potential yield. Fertility stress from too little K can be avoided.

Soil fertility is a factor that can be controlled. When inadequate fertility becomes adequate fertility, it helps the crop overcome low fertility stress and other stresses.

K x N
N rate, lb/A
Hay yield, t/A
3.5 3.0 2.5 2.0 1.5 1.0
160
240
60
No N
0 50 100 150
K fertilizer, lb K_2O/A

Figure 5-2. Effect of applied N and K on total forage yield of timothy (average of two years). Forage yields expressed as hay at 18% moisture content. (Quebec)

Review Questions:

4. (T or F) Potassium forms several compounds in plants.

5. (T or F) Potassium improves winter hardiness and increases disease resistance in plants.

6. A major role of K in crop growth is increased ______ use efficiency through its regulation of the opening and closing of ______, small pores on the underside of leaves.

7. (T or F) Potassium increases kernel weight but does not affect the number of kernels per ear of corn.

Plant Deficiency Symptoms

Potassium deficiency symptoms show up in many ways. One of the most common K hunger signs is scorching or firing along leaf margins (see **Appendix A**). Firing first appears on older leaves in most plants, especially grasses. Newer leaves will show hunger signs first on some plants and under some conditions...for example, high yielding cotton and soybeans in mid to late season.

Potassium deficient plants grow slowly. They have poorly developed root systems. Stalks are weak, and lodging is common. Seed and fruit are small and shriveled, and plants possess low resistance to disease. Some specific crop deficiency symptoms are listed below.

- Corn—firing or scorching on the outer edge (margin) of the leaf while the midrib remains green; unfilled, chaffy ears.

- Soybeans—firing or scorching begins on the outer edge of the leaf, edges becoming broken and ragged as the leaf dies; shrivelled, non-uniform seeds.

- Alfalfa—small white or yellowish dots around outer edges of leaves; leaf then turns yellow, and tissue dies.

- Fruit tree crops—yellowish green leaves curl upward along margins; scorched areas develop along edges which become ragged; small fruit which drop prematurely; poor storage, shipping, and canning qualities.

- Potatoes—upper leaves usually smaller, crinkled, and darker green than normal.

Potassium and Functional Foods

Functional foods, also called designer foods or pharma-foods, are associated with the prevention and treatment of at least four of the leading causes of death: cancer, diabetes, hypertension, and heart disease. They contain bio-active ingredients thought to enhance health or fitness. These ingredients—phytochemicals or nutraceuticals—are not among the traditional nutrients such as carbohydrates, fats, and proteins. Two of the phytochemicals contained in plants are isoflavones in soybeans and lycopene in tomatoes.

It has been demonstrated time and again that crop yield and quality can be manipulated by properly managing N, P, K, and other nutrients. For example, in an Ontario study, results

Table 5-6. **Band-applied K fertilizer boosted soybean yield and isoflavone content in a field at Paris, Ontario. Mean of two years and three soybean row widths (7.5, 15, and 30 in.).**

K fertilizer rate	Isoflavones, ppm[1]			Total	Yield, bu/A
	Genistein	Daidzein	Glycitein		
90 lb K_2O/A banded	688	579	122	1,389	37.4
No K	537	499	109	1,145	31.8
Difference, %	28	16	12	21	17

[1]ppm = parts per million

showed that when soybean yields responded to K fertilization, isoflavone levels increased, **Table 5-6**. Isoflavone content was also related to seed K concentration. It is not known if K plays a specific role in isoflavone synthesis, but it may since K is an important enzyme cofactor for many plant metabolic processes.

Recent studies have also demonstrated that most of the carotenoids, including lycopene, increased in tomatoes grown in sand culture as K content in the nutrient solution was increased from 0 to 260 parts per million (ppm). Lycopene increased by 73% over this K nutrient range. Undoubtedly, the importance of K fertilization will increase as science continues to relate its effects on crop yield and quality.

Forms of Potassium in the Soil

Although most soils contain huge amounts of K, frequently 20,000 lb/A or more, just a small percentage is available to plants over the growing season, often less than 2%. Soil K exists in three forms: Unavailable, slowly available, and available.

1. **Unavailable K**–Unavailable K is found in minerals (rocks). The K is released as soil minerals are weathered, but so slowly as to be unavailable to growing plants in a particular crop year. The weathering process is so slow, in fact, it could take hundreds of years to add significant amounts of available K to the soil. Generally, soils in the eastern parts of the U.S. are more highly weathered than those in the Midwest and West. Less weathered soils are richer in K than those which have undergone more extensive weathering.

2. **Slowly available K**–Slowly available K is "fixed" or trapped between layers of certain soil clays. Such clays shrink and swell during dry and wet soil conditions. Potassium ions can be trapped between these clay layers, becoming unavailable or only slowly released. Highly weathered soils don't contain many of the K-fixing clays. Sandy soils contain lesser reservoirs of slowly available K than do those containing greater amounts of clay.

3. **Available soil K**–Readily available K is made up of the K found in the soil solution plus the K held in exchangeable form by soil organic matter and clays.

Chapter 1 discussed cation exchange capacity (CEC). Remember that soil colloids have negative charges and attract cations, such as K^+. Soil colloids repel anions, such as nitrate (NO_3^-). So, cations are held in exchangeable form (adsorbed). These exchangeable cations are in equilibrium with those in the soil solution. This equilibrium may be represented as follows:

$$\text{Exchangeble K} \quad \longleftrightarrow \quad \text{Solution K}$$

Most soils contain 10 lb/A or less of solution K. This will supply an actively growing crop barely a day or two. But as the crop removes solution K, some of the exchangeable K moves into solution. It is replaced on the soil colloid by some other cation. This exchange continues until a new equilibrium is established. So, through the cation exchange process, K is continuously available for plant growth, if the soil contains enough K at the beginning of the growing season to supply the crop's needs.

Note: Some K can be exchanged directly from soil colloid to plant root when the two come in contact.

Fertilizer Potassium in the Soil

In the preceding section, two kinds of soil K were identified as being readily available for crop growth. They are solution, or soluble K, and exchangeable K.

- Solution K is found in soil water

- Exchangeable K is loosely held in an exchangeable form by soil clays and organic matter (colloids).

Plant roots can take up either form of available K for use in meeting their needs. The question is: "What form does fertilizer K take when it is applied to the soil and dissolved in the soil solution?"

The K in different nutrient sources (commercial fertilizers, organic materials, crop residues, cover crops, etc.) takes on the ionic form (K^+) when it dissolves. Thus, the K from all sources is the same. The following examples illustrate.

- Muriate of potash (KCl): $KCl \xrightarrow{Moisture} K^+ + Cl^-$

- Sulfate of potash (K_2SO_4): $K_2SO_4 \xrightarrow{Moisture} 2K^+ + SO_4^{2-}$

- Potassium nitrate (KNO_3): $KNO_3 \xrightarrow{Moisture} K^+ + NO_3^-$

- Organic materials $\xrightarrow[Mineralization]{Moisture}$ K^+ + accompanying anion(s); electrical balanced maintained

Once the K ionizes to the K^+ form, it makes no difference what the original source might have been. All the K^+ is now the same and is subject to the same fate in the soil. One or more of several things can happen to it.

- It can be attracted to the surfaces of soil clays and organic matter (CEC) and held in exchangeable form until it is taken up by a plant root or replaced on the exchange site by another cation.

- Some will remain in the soil solution.

- Some can be taken up immediately by the growing crop.

- Part of it can be leached. In sandy soils, some K can be leached out of the root zone. Because organic matter only weakly attracts K^+ on its CEC sites, K can be leached from organic soils.

- Part of it can be "fixed" (converted to an unavailable or slowly available form) in some soils. **See Production Concept 5-2.** Fixed K is only slowly available. The fixing capacity of some soils is why soil tests do not always reflect fertilizer applications and their effects on increasing soil test values. Fixed K is not measured in most soil tests.

The Potassium Cycle

The soil K cycle is shown in **Figure 5-4**. Like P, soil K inputs include manures, biosolids, plant residues, and the weathering of soil nutrients. Unlike N and P, however, K exists in various organic sources as inorganic K, rather than as a structural component of organic compounds.

Figure 5-4. The soil K cycle.

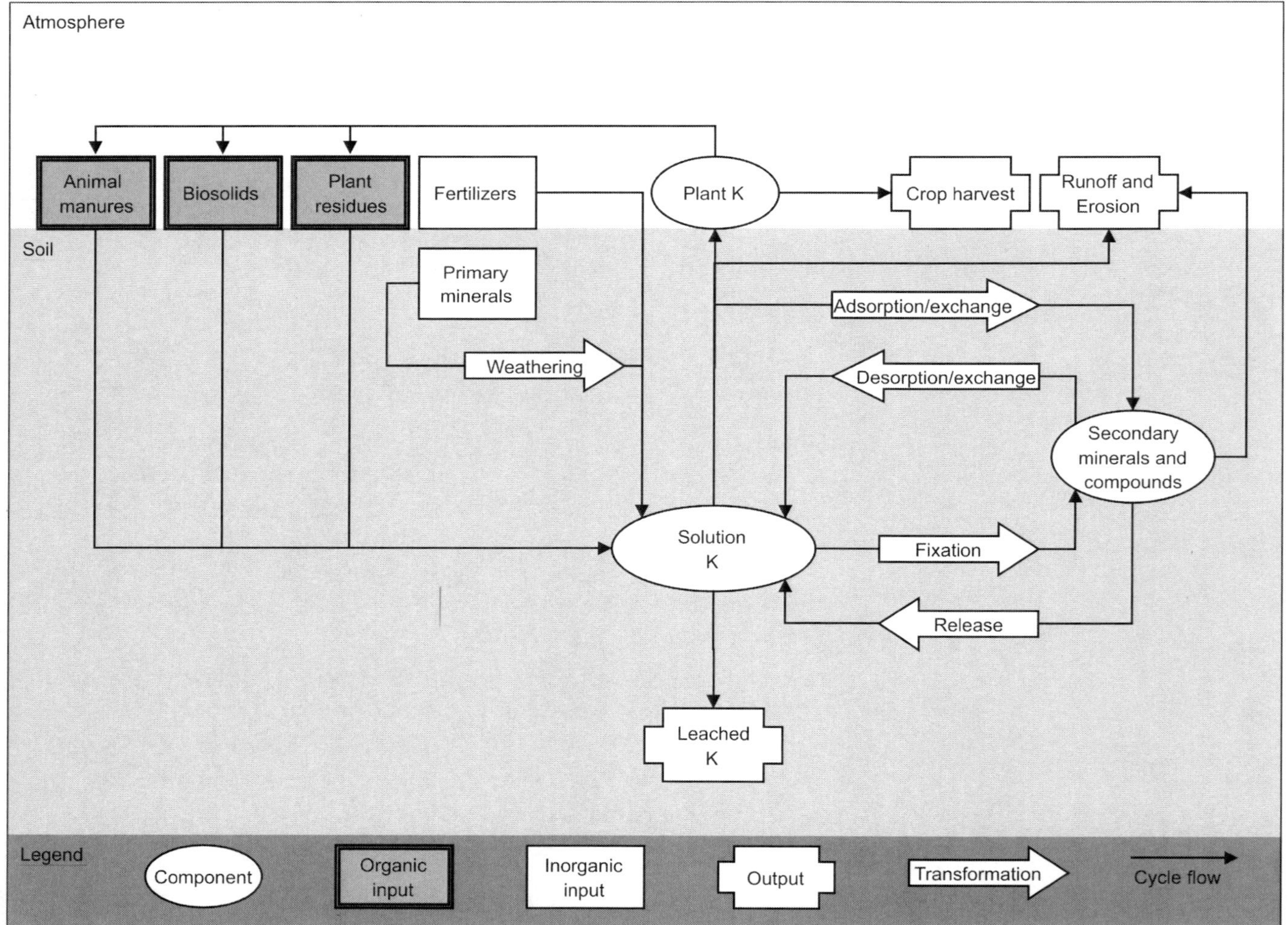

How Potassium Moves in the Soil

It is vital to maintain adequate K fertility levels in the soil because soil K does not move much, except in sandy or organic soils. Unlike N and some other nutrients, K tends to remain where fertilization puts it. When K does move, it is mainly by diffusion and only slowly...short distances through water films surrounding soil particles. Dry conditions and low soil temperatures slow this movement even more. High soil K levels speed it up.

Crop roots usually contact less than 3% of the soil in which they grow. So the soil must be well supplied with K to insure that plant needs at every stage of growth are met until harvest. Total root mass of corn occupies less than 1% of the soil volume, illustrating the importance of K diffusion to absorbing roots.

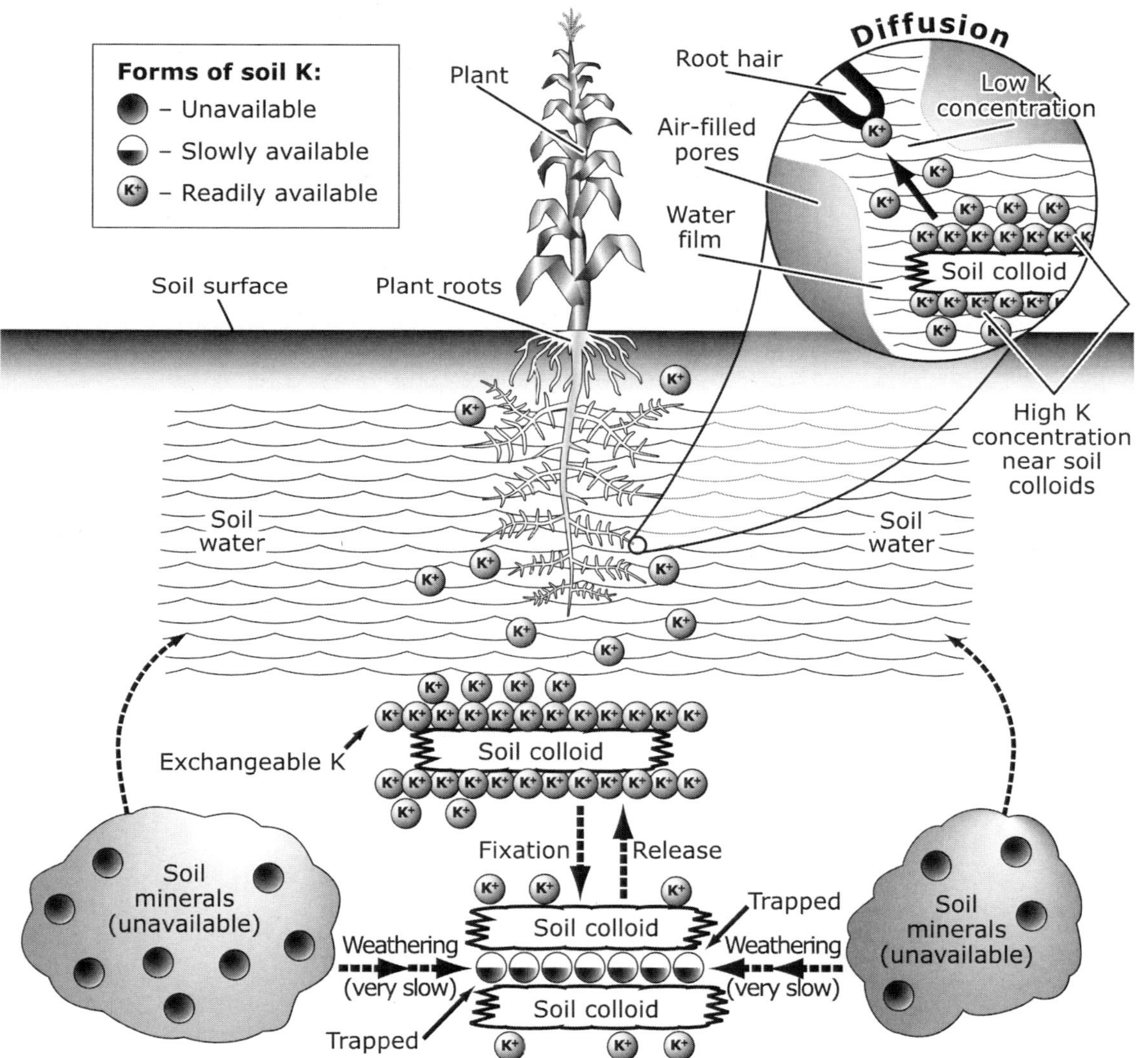

There are three forms of soil K—**unavailable, slowly available**, and **readily available**. This illustration shows how they relate to each other and to plant availability. Potassium moves to plant roots by diffusion, a slow process.

Soil Factors Affecting Potassium Uptake by Plants

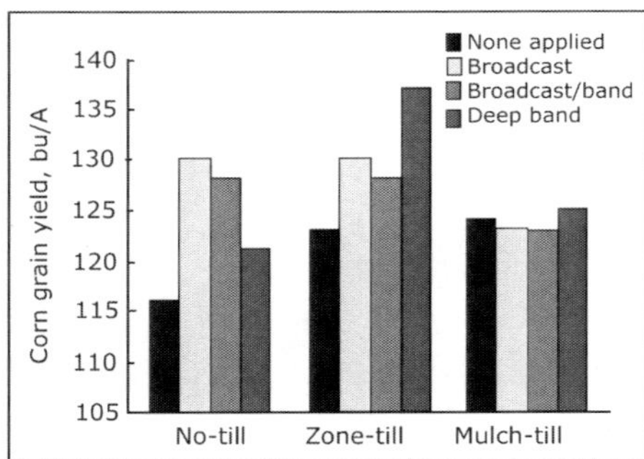

Figure 5-5. Corn yield response to K application method in three spring tillage systems. Average of two years at Paris, Ontario.

Review Questions:
23. (T or F) Fertilizer K can be taken up almost immediately by the growing crop once it is mixed with the soil.
24. Potassium is considered to be an immobile soil nutrient, but can be leached in ______ or ______ soils.
25. Potassium uptake by plant roots is affected by several soil factors, including ______, ______, ______ and ______.

Potassium is relatively immobile in the soil, reaching plant roots largely by diffusion. Therefore, any factor restricting root growth can decrease K uptake. Several of those factors are discussed below.

- Soil aeration—Uptake of K is affected more by poor aeration than are most other nutrients. Minimum tillage systems and compaction both limit K uptake and increase deficiency problems, largely because of reduced aeration and root growth. **Figure 5-5** shows the effects of three tillage systems on corn response to K.

- Soil test K—As the level of soil test K drops, root uptake decreases.

- Fixation—Soils which trap K and hold it in unavailable form reduce the amounts available for plant uptake.

- CEC—In general, soils with high CECs have greater storage capacity and supplying power for K.

- Soil temperature—Low soil temperatures reduce the availability and uptake of K by plant roots. Their effects can be partly offset by increasing soil K levels.

- Soil moisture—Moisture is needed for K to move...by diffusion...to plant roots for uptake. Both drought stress and excess moisture reduce the uptake of K.

The soil itself—its overall characteristics—determines how efficiently crops are able to take up and use K. These characteristics include parent material from which the soil was formed, amount and types of clay minerals it contains, vegetation under which it was formed, topography, drainage, depth...and so on. The grower must learn to manage his or her soil resource, adjust practices to optimize its productivity potential (including K use efficiency), and improve it where possible.

Methods of Applying Potassium Fertilizers

There is no "best" way to apply K. Methods depend on many soil and crop conditions...and on other management practices. Some factors which influence placement method or methods include:

- Crop,
- Available labor and equipment,
- Soil fertility level,
- Soil type,
- Fertilizer rate and time of application,
- Use of crop protection chemicals in combination with fertilizer, and
- Soil temperature.

A number of K application methods have been tested by both

researchers and farmers. There are several variations of these methods, including:

- Surface broadcast without incorporation,
- Broadcast and disked,
- Broadcast and plowed down,
- Banding, including various combinations of distances below and to the side of the seed,
- Plow sole placement,
- Deep placement or knifed,
- Surface strips,
- Fertigation, and
- Combinations of the above methods.

All these methods can be considered as variations between the two extremes...banding (row placement) with high K concentrations, but with minimum soil contact, and broadcasting with incorporation into the tillage layer.

Row applications concentrate nutrients for rapid early growth. This is important for young plants with limited root systems, particularly in cold and/or compacted soils. But too much fertilizer too close to seed can reduce germination and injure roots due to high salt concentrations. Row K should be placed beside and below seed level to reduce potential damage.

Broadcasting before planting may be the most convenient way to apply high amounts of K and other nutrients. When soil fertility levels are adequate, it is as efficient as row application. Yet, some soils can "fix" or make significant amounts unavailable. This, of course, reduces the immediate efficiency of broadcasting.

Fall banding of K into the ridge for ridge-till corn has been demonstrated to be an effective means of increasing K uptake and increasing yields. Though the specific cause of increased K deficiency in ridge-till corn, even at high K soil tests, is not specifically known, farmers must be aware of the potential problem and its solution for optimum yields and profits.

Combining row with broadcast is often the best way to apply fertilizer. It gives a fast, early start and a nutrient reservoir throughout the growing season. In general, crop responses to different methods of K placement are not as large nor as consistent as for N and P. However, cold, compacted, or dry soil conditions tend to place more stress on K uptake and may warrant placement of high concentrations of K in the vicinity of developing root systems, as shown by the results of an Ontario corn study, **Figure 5-6**.

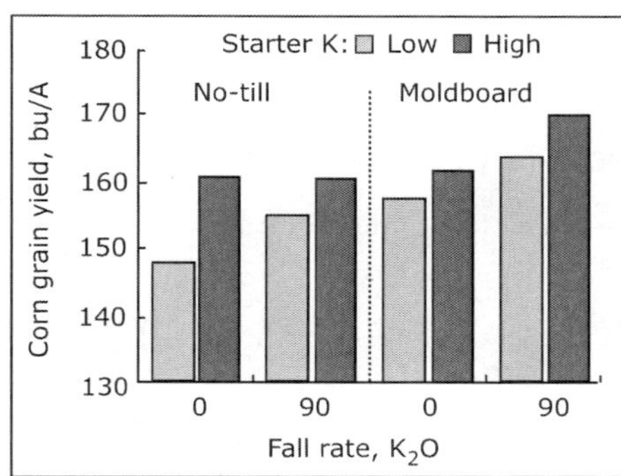

Figure 5-6. Corn grain yield response to K applied in the fall or as a spring starter at Kirkton, Ontario. Average of three years.

Review Questions:

26. (T or F) Soil type and soil fertility level influence placement methods of fertilizer K.

27. One advantage of row application is that a high ______ ______ is available for rapid early growth.

28. Row application should be made ______ and ______ seed level to reduce the potential for salt damage.

29. (T or F) Often, row and broadcast applications should be applied in combination.

North America has the world's largest known reserves of K, due mainly to the huge deposits of K ore in Canada. Most of these deposits are located in Saskatchewan, with additional reserves located in New Brunswick. Production in the U.S. is located mainly in the states of New Mexico (underground mines) and Utah (brine). There is also a small solution mine in Michigan.

Elemental K is not found in nature because of its chemical reactivity. Potash deposits occur as beds of solid salts beneath the Earth's surface and brines in dying lakes and seas. Potash is mined four major ways:

- **Conventional underground coal mining method.** This undercuts the face, drills, and blasts.

- **Continuous mining method.** This uses specially developed mining machines that take the ore directly from the vein, without blasting.

- **Solution mining.** A hot salt brine is pumped down to the potash bed, dissolving mainly K salts and returning the potash brine to the surface for refining.

- **Surface brine recovery.** Potassium (primarily as K sulfate, K_2SO_4) and other minerals are harvested by solar evaporation from natural surface brines in the Great Salt Lake in the U.S. and the Dead Sea in Israel and Jordan.

Potassium is mined from a number of minerals. Sylvinite and langbeinite are the most important.

- **Sylvinite** is composed primarily of KCl and sodium chloride (NaCl), containing 20 to 30% K_2O.

- **Langbeinite** is composed largely of K_2SO_4 and magnesium sulfate ($MgSO_4$), containing about 23% K_2O. (Brines containing K are about two-thirds water and contain about 3% K_2O.)

Potassium chloride, or muriate of potash, accounts for more than 90% of the K sold in the U.S. and Canada. It is water soluble and contains 60 to 62% K_2O. Most North American KCl is produced from sylvinite, but some comes from brines. The raw, impure ore is refined to fertilizer by crystallization or flotation processes. Most agricultural KCl is produced by the flotation process.

Fertilizer grade KCl is available in five particle sizes: 1) white soluble, 2) special standard, 3) standard, 4) coarse and 5) granular. Granular is well suited to bulk blending. The white soluble grade is ideal for clear liquids.

Potassium sulfate, also called sulfate of potash (SOP), contains about 50% K_2O and 18% sulfur (S). Because its chloride (Cl) content is below 2.5%, it is used for Cl-sensitive crops such as tree fruits and tobacco, and to supply S. It accounts for about 6% of total agricultural K sales. Potassium sulfate can be used where Cl buildup in soil becomes a problem.

Material	Percent composition				
	K_2O	Mg	S	N	Cl
KCl	60-62	-	-	-	45-47
K_2SO_4	50	-	18	-	-
$K_2SO_4 \cdot 2MgSO_4$	22	11	22	-	-
KNO_3	44	-	-	13	-

Table 5-7 summarizes the composition of common K fertilizer sources.

Sulfate of potash-magnesia ($K_2SO_4 \cdot 2MgSO_4$) is also called potassium-magnesium sulfate, "Sul-Po-Mag", and "K-Mag." It contains about 22% K_2O, 11% magnesium (Mg), and 22% S. It occurs in nature as the mineral langbeinite, which is crushed to make the commercial fertilizer product, and may be further refined. It is a good source of water-soluble K and Mg and is very important where Mg and/or S is deficient.

Note: Some formulations of K_2SO_4 and $K_2SO_4 \cdot 2MgSO_4$ are approved for "organic" crop production. The USDA Organic Standards permit use of potash under certain conditions as well.

Potassium nitrate contains little or no Cl or S. It contains about 44% K_2O and 13% N. Potassium nitrate is widely used as a source for foliar sprays of K to fruit, vegetables, and cotton. <SFM>

Chapter 6

The Secondary Nutrients

The secondary nutrients...calcium (Ca), magnesium (Mg), and sulfur (S)...are as important to plant nutrition as the primary nutrients. Calcium stimulates root and leaf development, and affects uptake and activity of other nutrients. Deficiencies of Ca don't usually occur in field conditions, but can appear in some situations. Magnesium is the central atom in the chlorophyll molecule, so it is actively involved in photosynthesis. The ratio of Mg to potassium (K) can be an important factor related to grass tetany. Sulfur has several key functions in plants, and the need for S is closely related to the amounts of nitrogen (N) available to crop plants. Deficiency symptoms may be similar to those of N, although S is not mobile in plants.

Essential Plant Nutrients

While Ca, Mg, and S are called secondary nutrients, that does not mean they play a secondary role in plant growth. They are as important to plant nutrition as primary nutrients, though plants don't usually require as much of them. Many crops contain as much S as phosphorus (P), sometimes more. Secondary nutrient deficiencies can also depress plant growth as much as primary nutrient deficiencies do. **Table 6-1** shows the amounts of Ca, Mg, and S contained in some crops.

Table 6-1. Calcium, Mg, and S taken up by some common crops.

Crop	Yield level	Ca[1]	Mg	S
Alfalfa	8 tons	175	40	04
Coastal bermudagrass	8 tons	52	26	44
Corn	160 bu	39	52	72
Cotton	1,000 lb lint	14	23	02
Grain sorghum	8,000 lb	60	40	93
Oranges	540 cwt	80	22	[2]
Peanuts	4,000 lb	20	25	12
Rice	7,000 lb	20	14	21
Soybeans	60 bu	26	24	02
Tomatoes	40 tons	30	36	45
Wheat	60 bu	16	18	51

[1]Estimated   [2]Data not available

Review Questions:

1. (T or F) Primary plant nutrients are more important in plant growth than the secondary nutrients.

2. Most plants contain (more, less, about the same) Mg than S; (more, less, about the same) S compared to P.

Calcium:
Role of Calcium in Plants

Calcium is taken up by plants as the Ca^{2+} cation. Once inside the plant, Ca functions in several ways, including the following.

- Stimulates root and leaf development.

- Forms compounds which are part of cell walls. This strengthens plant structure.

- Helps reduce nitrate-N (NO_3^--N) in the plant.

- Helps activate several plant enzyme systems.

- Helps balance organic acids in the plant.

- Is essential for nut development in peanuts.

- Some sources influence yields indirectly by reducing soil acidity (i.e. aglime, Ca carbonate [$CaCO_3$]). This lowers solubility and toxicity of manganese (Mn), copper (Cu), and aluminum (Al).

- Helps yields indirectly by improving root growth conditions and stimulating microbial activity, molybdenum (Mo) availability, and uptake of other nutrients.

Plant Deficiency Symptoms

Abnormal development of growing points—terminal buds— and poor root growth are common symptoms of a Ca deficiency. Young leaves and other new tissue develop symptoms first because Ca is not translocated within the plant. New tissue needs Ca pectate for cell wall formation, so a Ca deficiency can cause gelatinous leaf tips and growing points. In severe cases, the growing point dies, and roots turn black and rot.

Calcium deficiency can result in the foliage taking on an abnormal dark green color. Deficient plants might shed blossoms and buds prematurely. Tip burn on the young leaves of vegetable crops such as lettuce and celery can occur. Other symptoms include bitter pit or cork spot of apples and pears and blossom-end rot (see **Appendix A**) of tomatoes, peppers, and melons.

It is generally assumed that Ca is so abundant in soils that deficiencies don't occur under field conditions. In most cases this is true, particularly in acid soils. Usually the effects of high soil acidity limit growth before Ca becomes deficient. On alkaline soils, though, where available Ca can be low, fertilizer Ca may be required to supply plant requirements. In the western U.S., for example, foliar Ca is often applied to some celery varieties to prevent black heart, a disorder of young leaves. Also, Ca sprays and/or dips are used on apples, pears, and cherries to reduce fruit disorders such as cork spot, bitter pit, and cracking.

Calcium in the Soil

Total amounts of Ca in the soil range from less than 0.1% to as much as 25%. Arid, calcareous soils contain the highest levels. Newly drained organic soils often contain very little Ca and have extremely low pH values. Clay soils usually contain more Ca than sandy soils.

Since Ca exists as a cation, it is governed by cation exchange phenomena just as other cations are and is held as exchangeable Ca^{2+} on the negatively charged surfaces of soil clay and organic matter. It is usually the most dominant cation in the soil, even at low pHs, and normally occupies 70 to 90% of the sites of the soil's cation exchange complex. Like other cations, Ca is also present in the soil solution. It is part of the structure of several soil minerals. In fact, such soil minerals as dolomite, calcite, apatite, and Ca feldspars are a major soil source of Ca.

Soil Fertility Manual

Sources of Calcium

Calcium can be supplied in several ways. Because most Ca deficient soils are acid, a good liming program can add Ca most efficiently. Both calcitic and dolomitic limestones are excellent sources. Gypsum can also supply Ca when soil pH is high enough not to need aglime, but (in the unlikely event) where Ca might be deficient. Normal superphosphate...which is 50% gypsum...and to a lesser extent, triple superphosphate, can also add Ca to the soil. Some common sources of Ca are shown in **Table 6-2**.

Table 6-2. Common Ca sources.

Material	Ca, %	Relative neutralizing value, %[1]
Calcitic limestone	32	85-100
Dolomitic limestone	22	95-108
Basic slag	29	50-70
Gypsum	22	None
Marl	24	15-85
Hydrated lime	46	120-135
Burned lime	60	150-175

[1]Based on pure calcium carbonate at 100%.

When using Ca sources other than ground calcitic and dolomitic limestone, apply with caution. Excess hydrated lime and burned lime can partially sterilize the soil. Adding large amounts of Ca and Mg to K-deficient soils or adding Ca to an Mg-deficient soil can cause nutrient imbalances and poor crop growth. Supply all needed nutrients to alleviate growth-limiting conditions.

Magnesium:
Role of Magnesium in Plants

Magnesium is taken up by plants as the Mg^{2+} cation. Once inside the plant, it serves many functions. Magnesium is the central atom in the chlorophyll molecule, so it is actively involved in photosynthesis (**Figure 6-1**). Magnesium and N are the only soil nutrients that are constituents of chlorophyll. Much of the Mg in plants is found in the chlorophyll. Seeds are also relatively high in Mg, though grain crops such as corn have low levels in the seed. Magnesium also aids in phosphate metabolism, plant respiration, and the activation of many enzyme systems.

Figure 6-1. Hidden in the heart of each molecule of chlorophyll is an atom of Mg. Deprive a plant of Mg, and its chlorophyll molecules (and plant life) cease to exist.

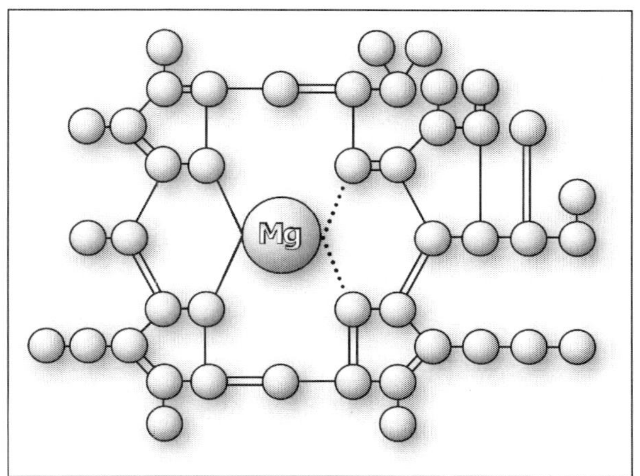

Plant Deficiency Symptoms

Magnesium deficiency symptoms first appear on lower (older) leaves, generally, because Mg is translocated within the plant (see **Appendix A**). They show a yellowish, bronze, or reddish color, while leaf veins remain green. Corn leaves are yellow-striped, with green veins. Crops such as potatoes, tomatoes, soybeans, and cabbage also show the orange-yellow color and green veins.

Imbalance between Ca and Mg in low cation exchange capacity (CEC) soils may accentuate Mg deficiency. When the Ca to Mg ratio becomes too high in these soils, plants may take up less Mg. This can occur when a grower has limed with only calcitic aglime for several years on soils relatively low in Mg. Magnesium deficiency may also be accentuated by high K rates or by high availability of ammonium-N (NH_4^+–N) when the soil contains borderline Mg.

Magnesium in the Soil

Soil Mg, other than that added in fertilizer or liming materials, comes from the weathering of rocks containing such minerals as biotite, hornblende, dolomite, and chlorite. Being a cation, Mg^{2+} is subject to cation exchange. It is found in the soil solution and is adsorbed to clay and organic matter surfaces. Soils usually contain less Mg than Ca because Mg is not adsorbed as tightly by clay and organic matter and is more subject to leaching. Also, most parent materials contain less Mg than Ca. Although many soils contain enough Mg to support plant growth, Mg deficiencies frequently occur, most often on coarse-textured, acid soils developed under high rainfall conditions. Deficiencies can also develop on calcareous soils where irrigation water contains high bicarbonate or on sodic soils.

The ratio of Mg to K can be an important factor under some conditions. For example, fertilizing with K can reduce the uptake of Mg by grasses being grazed by livestock, resulting in low blood serum Mg and a condition known as grass tetany. Low soil temperatures and plenty of soil moisture in the presence of only moderate amounts of K result in higher K uptake, compared to Mg, and the development of tetany-prone grass forages. High levels of plant-available P are important in Mg uptake and lowering of grass tetany potential in forage grasses.

Sources of Magnesium

The most common source of Mg is dolomitic limestone, an excellent material which provides both Ca and Mg while neutralizing soil acidity. Other sources include potassium-magnesium sulfate ($K_2SO_4 \bullet 2MgSO_4$), also known as K-Mag or Sul-Po-Mag, Mg sulfate ($MgSO_4$), Mg chloride ($MgCl_2$), Mg oxide (MgO), and basic slag.

Table 6-3 shows common Mg sources and the percent Mg they contain. The sulfate (SO_4^{2-}) forms of Mg are more soluble than dolomitic aglime and may be the preferred Mg source on those soils where a quick crop response is required.

Table 6-3. Common Mg sources.

Material	Mg, %
Dolomitic limestone	3-12
Magnesium oxide (magnesia)	55-60
Basic slag	3
Magnesium sulfate	9-20
Potassium-magnesium sulfate	11
Magnesium chloride (solution)	7.5

Sulfur:
Role of Sulfur in Plants

Unlike Ca and Mg, which are taken up by plants as cations, S is absorbed primarily as the SO_4^{2-} anion. It can also enter plant leaves from the air as S dioxide (SO_2) gas. Sulfur is a part of every living cell and is a constituent of two of the 20 amino acids which form proteins. Other functions of S in the plant:

- Helps develop enzymes and vitamins.

- Promotes nodulation for N fixation by legumes.

- Aids in seed production.

- Is necessary in chlorophyll formation, although it is not a constituent of chlorophyll.

- Is present in several organic compounds which give the characteristic odors to garlic, mustard, and onion.

Plant Deficiency Symptoms

Plants deficient in S show a pale green coloring of the younger leaves, although the entire plant can be pale green and stunted in severe cases (see **Appendix A**). Leaves tend to shrivel as the deficiency progresses.

Sulfur, like N, is a constituent of proteins, so deficiency symptoms are similar to those of N. Nitrogen deficiency symptoms are more severe on older leaves, however, because N is a mobile plant nutrient and moves to new growth. Sulfur, on the other hand, is immobile in the plant, so new growth suffers first when S levels are not adequate to meet crop need. This difference is important in distinguishing between N and S deficiencies, particularly in early stages.

Plants deficient in S can be thin-stemmed and spindly. Crops such as cabbage and canola may develop a reddish color, first appearing on the underside of leaves and on stems. On alfalfa, leaves are long and slender, and branching is reduced.

The need for S is closely related to the amounts of N available to crop plants. This close relationship should not be surprising, since both are constituents of proteins and are associated with chlorophyll formation. Data in **Table 6-4** show how best corn yields were obtained when both N and S were applied.

Table 6-4. Corn response is best when both N and S are applied.

S rate, lb/A	N rate, lb/A			
	0	75	150	Average
	- - - - - - - - - Yield, bu/A - - - - - - - - -			
0	63	128	145	112
10	79	143	153	125
20	92	146	155	131

Minnesota

Nitrogen and S are further linked by the role of S in the activation of the enzyme NO_3^- reductase, which is necessary for the conversion of NO_3^- to amino acids in plants. Low activity of this enzyme depresses soluble protein levels, while raising NO_3^- concentrations in plant tissue.

High levels of NO_3^-, which accumulate when S is deficient, drastically inhibit seed formation in sensitive crops such as canola. **Table 6.5** shows how imbalance of N and S can affect canola yields. Nitrate can also be toxic to animals consuming forages deficient in S. Adequate levels of S improve Mg utilization by ruminants by reducing non-protein N levels (NO_3^-).

Table 6-5. Nitrogen and S imbalance affect canola yield in Canadian Prairies.

Fertilizer applied	Canola seed yield, bu/A		
	Alberta	Saskatchewan	Manitoba
No fertilizer	7	13	18
N P K	2	12	5
N P K S	35	33	38

Scientists have often suggested that the N:S ratio (total N to total S) in plants is a good diagnostic guide for determining S deficiency. Ratios of 10:1, 15:1, 7:1, 11:1, and others have been considered. Whether or not such ratios are valid, there is a strong relationship between N and S, one that cannot be ignored when N fertilizer use efficiency is evaluated. **Table 6-6** illustrates the point. Coastal bermudagrass responded to S fertilization. Sulfur also increased N use efficiency, improving profit potential and reducing the chance of NO_3^- leaching into groundwater.

Table 6-6. Sulfur boosts yield of Coastal bermudagrass and improves N uptake and recovery.

N rate, lb/A	Sulfur applied	Yield, tons/A	Nitrogen	
			Uptake, lb/A	Recovery, %[1]
0	No	2.4	81	—
	Yes	2.6	88	—
200	No	4.6	186	93
	Yes	5.2	223	112
400	No	5.1	236	59
	Yes	6.1	306	76

[1](N uptake ÷ N applied) x 100

Arkansas
Soil Fertility Manual

Sulfur in the Soil

Inorganic soil S...the form available to plants...occurs as the SO_4^{2-} anion. Because of its negative charge, SO_4^{2-} is not attracted to soil clay and organic matter surfaces except under certain conditions. It remains in the soil solution and moves with soil water, so it is readily leached. Certain soils accumulate SO_4^{2-} in the subsoil. This would be available to deep-rooted crops. In arid regions, sulfates of Ca, Mg, K, and sodium (Na) are the predominant inorganic S forms.

Much of the soil S in humid regions is associated with organic matter. Through biological transformations, similar to those of N, SO_4^{2-} and SO_4^{2-} compounds are produced that are available to the plant.

The S cycle, **Figure 6-2**, shows the relationships among atmospheric, fertilizer, and soil sources of S. Proper management insures efficient use of S, with minimal loss by leaching or erosion.

The number of soils deficient in S is increasing. There are several contributing factors, including:

- Increased crop yields that remove large amounts of S.

- Increased use of high analysis fertilizers containing little or no incidental S.

Figure 6-2. The sulfur cycle.

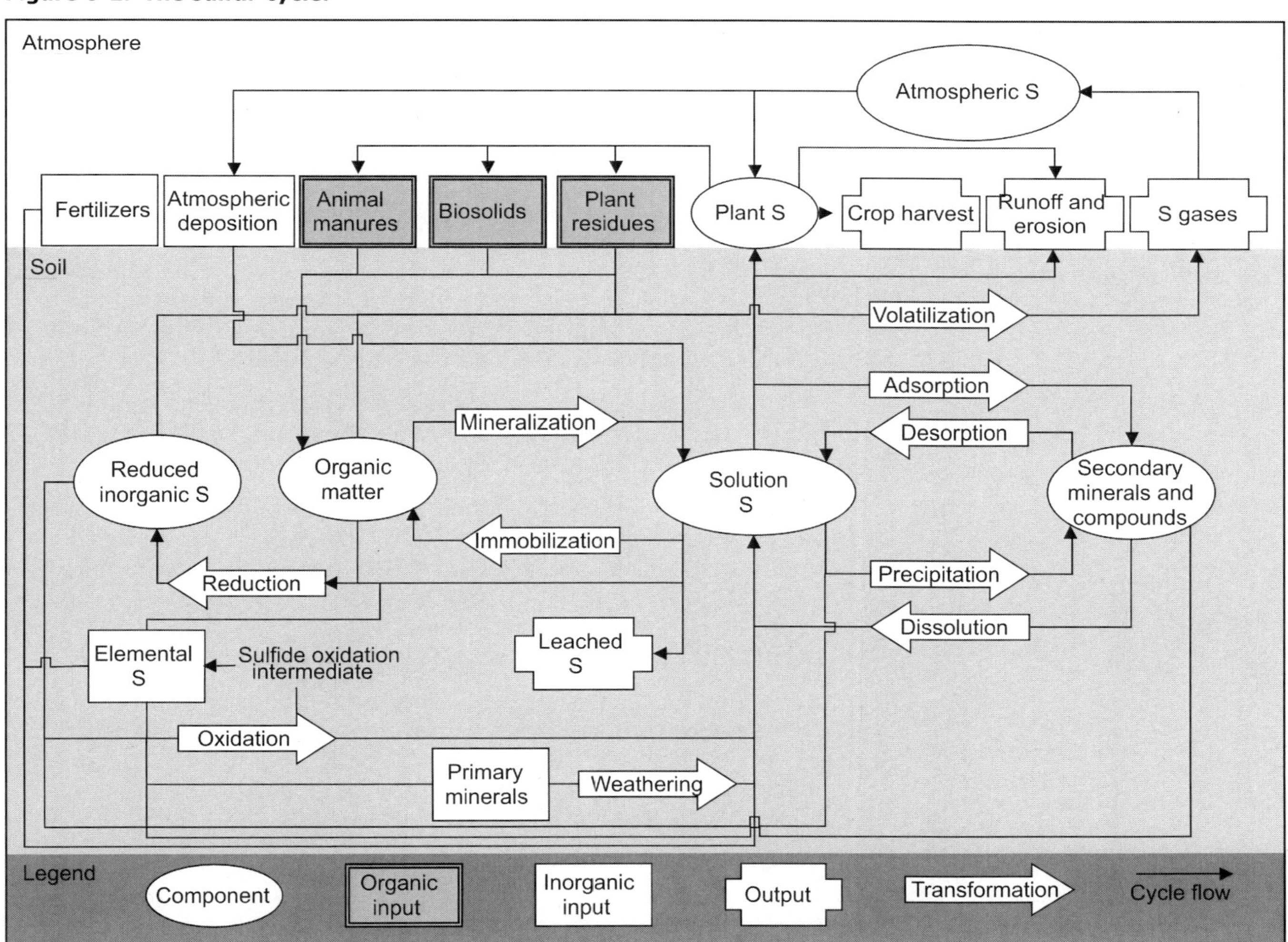

- Decreased use of high S fuels...and improved S removal techniques from stack gases.

- Less use of S-containing pesticides.

- Immobilzation of S in organic matter which accumulates because of conservation tillage practices.

- A greater awareness of S needs for profitable crop yields and quality.

Plant analysis and soil testing, including subsoil, are recommended for S on those soils suspected of being marginal or deficient. Other factors contribute to potential S deficiencies and should be considered when S recommendations are being made.

- Crop to be grown—High yielding forage crops such as the hybrid bermudagrasses and alfalfa remove more S and generally respond more frequently than most grain crops (see **Table 6-1**).

- Soil texture—Leaching of SO_4^{2-} from sandy soils is more likely than from finer-textured loams and clays. Crop response to S is most common on coarse-textured soils.

- Organic matter – Soils containing less than 2% organic matter are most commonly S deficient; however, deficiencies do occur on soils containing higher levels. Each percent organic matter releases about 5 lb S/A/yr.

- Irrigation water quality – Analyze water sources in order to determine their S concentrations.

Sources of Sulfur

Soil organic matter has already been mentioned as the primary soil S source. More than 95% of the S found in the soil is tied up in organic matter. Other natural sources include animal manures, irrigation water, and the atmosphere.

Animal manures contain S levels ranging from less than 0.02 to about 0.3%. Obviously, the content varies considerably, depending on species, method of storing and application, etc.

Sulfur dioxide and other atmospheric gases dissolved in rain and snow can contribute up to 20 lb S/A/year...even more in some industrialized areas. **Figure 6-3** shows an approximate view of sulfate ion concentrations in the U.S. recently. For a current look at trends, visit the website at **<http://nadp.sws.uiuc.edu/isopleths/maps2005/>**.

Irrigation water can contain fairly high levels of S. When the SO_4^{2-}-S content of irrigation water exceeds 5 parts per million (ppm), a S deficiency is highly unlikely. Still, starter fertilizer applications of S for some crops may be beneficial because of potential SO_4^{2-} leaching caused by rainfall or snowmelt.

Most S fertilizer sources are forms of SO_4^{2-} and are moderately to highly water soluble. Soluble forms also include bisulfites, thiosulfates, and polysulfides. The most important water

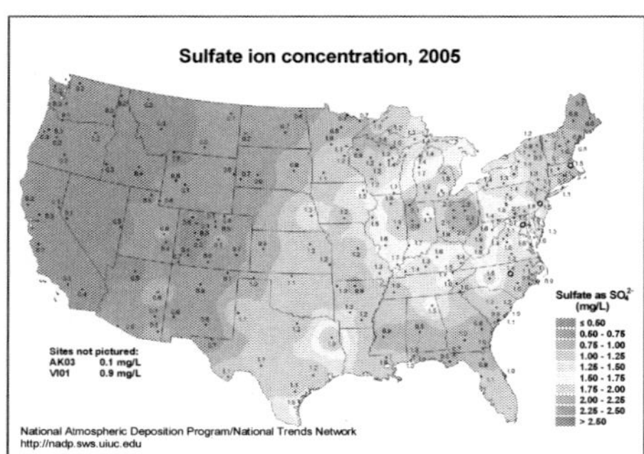

Figure 6-3. Sulfate ion concentrations in the U.S. (SO_4^{2-} mg/L). Adapted from National Atmospheric Deposition Program/ National Trends Network.

Soil Fertility Manual

insoluble S fertilizer is elemental S ($S°$), which must be oxidized to the SO_4^{2-} form before plants can use it. Bacterial oxidation of $S°$ in the soil is favored by

- Warm soil temperatures,

- Adequate soil moisture,

- Soil aeration, and

- Fine particle sizes.

Unless relatively high rates of $S°$ are applied (50 to 100 lb/A), it may take several weeks to months of warm, moist conditions for sufficient conversion to SO_4^{2-} to meet demands of high yielding crops.

Table 6-7 lists the common S fertilizer sources, along with their chemical formulae and S contents.

The water soluble SO_4^{2-} sources are immediately available to plants and should be used when S is needed quickly. They are commonly used in dry fertilizers, although $(NH_4)_2SO_4$ solutions are also common. Ammonium thiosulfate (ATS) is a liquid containing 12% N and 26% S, well-suited for use in fluid fertilizers or addition to irrigation water (fertigation). It should not be seed placed and, if band applied, should be at least 1 in. from the seed. Another liquid S source, K thiosulfate ($K_2S_2O_3$), contains 25% K_2O and 17% S and is also compatible in fluid fertilizer formulations and fertigation. Ammonium polysulfide (APS) is a red fluid S source with a strong odor of ammonia (NH_3), commonly applied in irrigation water. Sulfur in APS must be oxidized to the SO_4^{2-} form to be available to plants.

Table 6-7.   Common sulfur sources.

Fertilizer material	Chemical formula	Sulfur content, %
Ammonium sulfate	$(NH_4)_2SO_4$	24
Ammonium thiosulfate	$(NH_4)_2S_2O_3 \bullet 5H_2O$	26
Ammonium polysulfide	$(NH_4)_2S_2$	40-50
Potasssium sulfate	K_2SO_4	18
Potassium-magnesium sulfate	$K_2SO_4 \bullet 2MgSO_4$	22
Elemental sulfur	$S°$	>85
Gypsum	$CaSO_4 \bullet 2H_2O$	12-18
Magnesium sulfate	$MgSO_4 \bullet 7H_2O$	14
Potassium thiosulfate	$K_2S_2O_3$	17

Although $CaSO_4$ is less water soluble than the other SO_4^{2-} sources, it is effective and inexpensive.

Fertilization with $S°$ results in a slower crop response than SO_4^{2-} sources because it is water insoluble. To be effective, it should be incorporated into the soil well in advance of crop needs. Used properly, however, $S°$ is an agronomically effective and economically efficient S source in some soils.

One objection to the use of finely ground S is discomfort to the user. It is dusty...and can be a fire hazard under confined conditions. The problem is usually avoided by granulating the S with bentonite clay. ⟨SFM⟩

Chapter 7

The Micronutrients

Eight of the essential plant nutrients are called micronutrients. While micronutrients are not miracle workers, deficiencies can cause serious consequences in crops if not prevented or corrected. Soil pH and relative amounts of micronutrients in soils are important factors in the availability of micronutrients. Included in this chapter are sections describing the functions, sources, deficiency symptoms, possible toxicities, and management considerations for the micronutrients: boron (B), chloride (Cl), copper (Cu), iron (Fe), manganese (Mn), molybdenum (Mo), nickel (Ni), and zinc (Zn).

Micronutrients Are Essential for Plant Growth

Eight of the 17 essential plant nutrients are called micronutrients. They are B, Cl, Cu, Fe, Mn, Mo, Ni, and Zn.

They are as important to plant nutrition as primary and secondary nutrients, though plants don't require as much of them. A lack of any one of the micronutrients in the soil can limit plant growth, even when all other essential nutrients are present in adequate amounts.

The need for micronutrients has been known for many years, but their broad use in fertilizers is a fairly recent practice. Why have micronutrients become so important in recent years? Several reasons stand out:

• **Crop yields**—Increasing per-acre crop yields remove higher amounts of micronutrients. As more and more micronutrients are removed from the soil, some soils cannot release enough of them to meet today's high-yield crop demands.

• **Past fertilizer practices**—When crop yields were not so high and nitrogen (N), phosphorus (P), and potassium (K) fertilization was not as common as it is today, one of the three primary nutrients usually was the first limiting growth factor.

• **Fertilizer technology**—Today's high analysis production removes impurities much better than older manufacturing processes. Micronutrients are not commonly provided as "incidental" ingredients in fertilizers as they once were.

Review Questions:
1. The eight essential micronutrients are ______, ______, ______, ______, ______, ______, ______, and ______.
2. The chemical symbol for boron is ______; copper ______; zinc ______.
3. (T or F) High crop yields have influenced the increase in micronutrient deficiencies.

Micronutrients Are Not Miracle Workers

Micronutrients are not miracle workers, although a shortage of any one of them can limit growth and yields...can even cause plant death when deficient. To assign special value to micronutrients alone is incorrect and misleading.

Micronutrient fertilization should be treated as any other production input. If a micronutrient deficiency is suspected, it can be pinpointed through soil tests, plant analyses, and/or local field demonstrations. One should develop the habit of closely observing the growing crop for potential problem areas. Field diagnosis is one of the most effective tools available in production management.

Soil-Plant Relationships

Soils vary in micronutrient content, and they usually contain lower amounts of micronutrients than of primary and secondary nutrients.

Table 7-1 shows total micronutrients found in soils and how much 150 bu/A corn and 1,000 lb/A lint cotton yields can remove.

Table 7-1. Micronutrient contents in soils and crop removal.

Nutrient	Range in soils, total lb/A	Estimated crop removal, lb/A	
		Corn, 150 bu	Cotton, 1,000 lb lint
Boron	20-200	0.06	0.05
Copper	2-400	0.05	0.03
Iron	10,000-200,000	0.10	0.07
Manganese	100-10,000	0.08	0.30
Molybdenum	1-7	0.03	0.02
Zinc	20-600	0.15	0.06

Remember, total soil content of micronutrients does not indicate amounts available for plant growth during a single growing season. It does indicate relative abundance and potential supplying power for a particular nutrient.

Soil pH has a significant effect on micronutrient availability. Availability decreases as pH increases for all micronutrients except Mo and Cl. **Table 7-2** shows the soil pH range where each micronutrient is most available.

Table 7-2. Best pH ranges for micronutrient availability.

Micronutrient	Symbol	pH range for maximum availability
Boron	B	5.0-7.5
Chloride	Cl	Not affected
Copper	Cu	5.0-7.0
Iron	Fe	4.0-6.5
Manganese	Mn	5.0-6.5
Molybdenum	Mo	7.0-8.5
Zinc	Zn	5.0-7.0

In very low pH soils, some micronutrients can become soluble enough to be toxic to plants. Manganese, for example, can inhibit root growth in certain acid soils. Liming the soils to more neutral pH values reduces the danger of toxicity.

As pH values increase...by liming or naturally...chances of micronutrient deficiencies increase. Molybdenum and Cl⁻ are the exceptions. As pH increases, Mo availability goes up, while Cl⁻ availability is not affected. This is one reason liming acid soils can affect soybean yields as much or more than Mo seed treatment does.

Relative amounts of micronutrients in the soil, especially metals, determine their availability and may be more important than absolute amounts of each. This relationship can make soil test results misleading, unless the levels of other micronutrients (and primary and secondary nutrients) are considered.

In the following sections, each of the micronutrients is discussed individually.

Boron

Boron exists primarily in soil solution as the BO_3^{3-} anion. That is the form commonly taken up by plants. Boron deficiencies are widespread across North America and other parts of the world. Responses to B fertilization have been documented in most states in the U.S. and many areas in Canada. Alfalfa frequently responds to B, but responses also occur in a large number of fruit, vegetable, and field crops.

Boron is essential for germination of pollen grains and growth of pollen tubes and for seed and cell wall formation. It also forms sugar/borate complexes associated with sugar translocation and is important in protein formation.

Boron deficiency generally stunts plant growth, the growing point and younger leaves first. This indicates B is not readily translocated in the plant. Boron deficiencies include reduced flowering; thickened, curled, chlorotic leaves; and soft or necrotic spots in fruits or tubers (see **Appendix A**). Here are specific hunger signs:

Celery—crooked stem

Peanuts—hollow heart

Apples—corky core

Alfalfa—rosetting, yellow top, death of terminal bud

Table beets and sugarbeets—black heart

Cotton—ringed or banded leaf petioles, with dieback of terminal buds, causing rosetting effect at the top of the plant (seldom seen in the field); ruptured squares and thick, green leaves that stay green until frost and are difficult to defoliate.

Crops vary significantly in their responsiveness to B (**Table 7-3**). Most legumes, as well as several fruits and vegetables, are highly responsive to B. Other vegetables show somewhat less response. Grains are generally less responsive.

Table 7-3. Responsiveness of crops to B.

Most response	Medium response	Least response
Alfalfa	Broccoli	Beans
Cauliflower	Cabbage	Blueberries
Celery	Carrots	Cucumbers
Sugarbeets	Lettuce	Corn
Table beets	Spinach	Onions
Turnips	Sweet Corn	Potatoes
Peanuts	Tomatoes	Small grains
Cotton	Asparagus	Sorghum
Apples	Canola	Sudan grass
Clover	Radish	Soybeans

There are several factors that influence B availability in the soil:

- **Organic matter**—Organic matter is the most important soil source of B. In hot, dry weather, decomposition slows down in the soil surface where most of the organic matter is found. This can lead to a B deficiency. In cold weather, organic matter decomposition also slows, and low B release affects many cold crops (i.e., Brussels sprouts, radishes), and other early-planted species.

- **Weather conditions**—Dry weather restricts root activity in the surface soil and can cause temporary B deficiency. Symptoms may tend to disappear as soon as the surface soil receives rainfall. Root growth continues, but yield potential is often cut.

- **Soil pH**—Plant availability of B is good between soil pH values of 5.0 to 7.5, with maximum availability occurring between about 6.0 and 7.0. Liming acid soils can lower B availability and enhance response to B-containing fertilizers. In a Texas study, the use of finely ground limestone (more reactive) resulted in higher requirements for B for optimum rose clover yield because it raised soil pH more rapidly, reducing the solubility of B. Data in **Figure 7-1** show the interaction of two different grades of aglime on rose clover response to B. With the less reactive, coarse liming material, several factors could have influenced the drop in yield at the high B rate.

In a related study, the effects of liming acid soils on alfalfa response to B were evaluated. Estimated yields at varying levels of hot-water-soluble B and soil pH in the 2- to 6-in. depth, with B applied at 2 lb/A and soil Mn at 7.5 parts per million (ppm), are shown in **Table 7-4**. Yields were increased by increasing B levels at all pHs and, conversely, by increasing pH at all levels of soil B. Researchers estimated that 76% of the variability in alfalfa yield could be attributed to soil pH, soil B, B applied, and soil Mn.

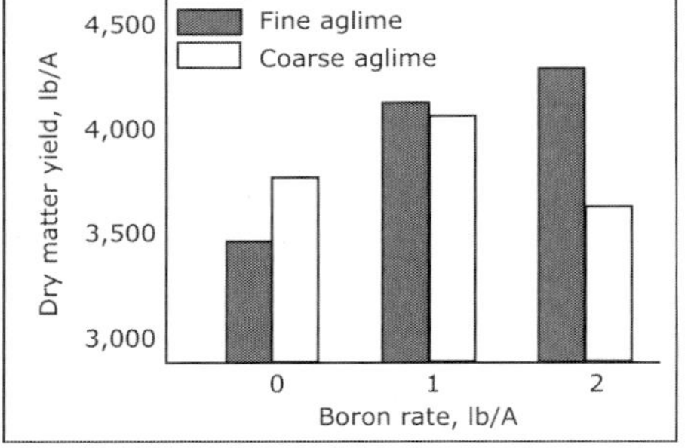

Figure 7-1. Rose clover response to aglime and B (Texas).

Table 7-4. Estimated response of alfalfa to soil pH and hot water-soluble B in the 2- to 6-in. depth of a Darco loamy fine sand with applied B at 2 lb/A and soil Mn at 7.5 ppm in 1994.

Soil pH	Soil B, ppm				
	0.3	0.4	0.5	0.6	0.7
	- - - - - - - - - - - - - Dry matter, t/A - - - - - - - - - - -				
5.7	0.97	1.17	1.90	3.15	4.94
6.2	2.05	2.24	2.98	4.23	6.01
6.7	2.93	3.14	3.86	5.16	6.91
7.2	3.64	3.84	4.57	5.82	7.60
7.7	4.16	4.36	5.08	6.34	8.13

- **Soil texture**—Coarse-textured (sandy) soils, which are composed largely of quartz, are typically low in minerals that contain B. Plants growing on such soils commonly show B deficiencies.

- **Leaching**—Boron is mobile in the soil and is subject to leaching. Leaching is of greater concern on sandy soils and/or in areas of high rainfall.

Crops vary widely in their needs for—and tolerance to—B. Yet, the line between deficient and toxic amounts is more narrow than for any other essential nutrient. So B should be used very carefully, especially in a rotation involving crops with different sensitivities to B.

Table 7-5 shows common B fertilizer sources, their B contents and water solubility.

Table 7-5. Boron fertilizer sources, B contents, and water solubility.

Source	B, %	Water soluble
Borax	11	Yes
Sodium pentaborate	18	Yes
Sodium tetraborate		
Fertilizer borate 46	14	Yes
Fertilizer borate 65	20	Yes
Boric acid	17	Yes
Colemanite	10	Low
Solubor	20	Yes

It is important that B fertilizers are applied uniformly because of the narrow range between deficiency and toxicity. Rates of B fertilization depend on several factors, including: soil tests, plant analysis, plant species, crop rotation, weather conditions, cultural practices, and soil organic matter.

Boron can be applied to the soil as a broadcast or band application or applied as a foliar spray or dust. Soil application rates for responsive crops may be as high as 3 lb B/A and for low and medium responsive crops, 0.5 to 1.0 lb B/A. **Table 7-6** shows how B reduces the incidence of barren stalks and increases corn yield.

Table 7-6. Boron reduces barren stalks and increases corn yield.

Boron rate, lb/A %	Barren stalks, bu/A	Yield,
0	23	148
1	27	161
2	19	170
4	18	164

Soil applied.   Florida

Chloride

Chloride is taken up by plants as the Cl⁻ anion. It is active in energy reactions in the plant. Specifically, it is involved in the chemical breakdown of water in the presence of sunlight and activates several enzyme systems. It is also involved in transporting several cations...K, Ca, Mg...within the plant and regulating the actions of stomatal guard cells, thus controlling water loss and moisture stress and maintaining turgor.

Research has shown that Cl⁻ diminishes the effects of fungal root diseases such as take-all and common root rot on small grains. It also helps suppress infections of small grain fungal leaf and head diseases. Lowered incidence of stalk rot in corn has been related to adequate Cl^-. It is speculated that Cl^- competes with nitrate (NO_3^-) uptake, tending to promote the use of ammonium (NH_4^+–N). Higher concentrations of NO_3^- in plants have been related to disease severity.

Chloride can be broadcast preplant or topdressed with N. Small grain research in Kansas and Oregon has not shown significant yield difference in timing of Cl^- applications. However, heavy winter rainfall may reduce carryover of Cl^- on sandy soils because of high Cl^- mobility. Where there is significant potential for Cl^- leaching during the growing season, a topdress application is recommended at the same time N is applied.

Data in **Table 7-7** show wheat responses to Cl^- on low testing soils.

Table 7-7.   Wheat responses to Cl^-.

Chloride rate, lb/A	Wheat yields, bu/A			
	Year 1	Year 2	Year 3	Year 4[1]
0	37	55	62	77
30	45	61	66	–
60	–	61	69	80
90	–	61	67	–
Soil test Cl^-:	–	low	low	med-high

[1]Average of six varieties   Kansas

About 30 to 60 lb Cl^-/A per surface 2 ft. of soil seems to be adequate for top yields of small grains. It can be provided by fertilizer or the soil. The most practical source is potassium chloride (KCl) which contains about 47% Cl. Preplant, at seeding, and top-dressed applications have all been effective. Higher rates should be applied preplant or topdress. Chloride is highly mobile in the soil and should be managed accordingly.

Chloride can have negative effects on crops such as tobacco, some soybean varieties, potatoes, and some tree crops. Effects vary with crop varieties or root stock and intended crop use.

Copper

Plants take up Cu as the Cu^+ and Cu^{2+} cations. Copper is necessary to chlorophyll formation in plants and catalyzes several other plant reactions, although it is not usually a part of the product(s) formed by those reactions. It may play a role in vitamin A production. A deficiency interferes with protein synthesis.

Common symptoms of Cu deficiency include dieback in citrus and blasting of onions and some other vegetable crops. Copper deficient small grains may not head or form grain. Many vegetable crops show Cu hunger, with leaves that lose turgor and develop a bluish-green shade before becoming chlorotic and curling. Plants fail to flower.

Organic soils are most likely to be Cu deficient. Such soils usually contain adequate levels of Cu, but hold it so tightly that only small amounts are available to the crop. Sandy soils, low in organic matter content, may also become deficient in Cu because of leaching losses. Heavy, clay-type soils are least likely to be Cu deficient.

Other metals in the soil...Fe, Mn, aluminum (Al)...affect the availability of Cu for plant growth. This effect is independent of soil type.

Like most other micronutrients, large quantities of Cu can be toxic to plants. Excessive amounts depress Fe activity and may cause Fe deficiency symptoms to appear in plants. Such toxicities are not common. **Table 7-8** shows common Cu sources, their Cu contents, water solubilities, and application methods.

Table 7-8. Copper fertilizer sources, Cu contents, water solubility, and application methods.

Source	Content Cu, %	Water soluble	Application methods
Copper sulfate	22.5	Yes	Foliar, Soil
Copper ammonium phosphate	30.0	Slight	Foliar, Soil
Copper chelates	Variable	yes	Foliar, Soil
Other organics	Variable	Yes	Foliar, Soil

Iron

Iron is taken up by plants as the ferrous (Fe^{2+}) cation. It is a catalyst to chlorophyll formation. It acts as an oxygen carrier in the nodules of legume roots, and also helps form certain respiratory enzyme systems. Iron deficiency shows up as a pale green leaf color (chlorosis), with sharp distinction between green veins and yellow interveinal tissues (see **Appendix A**).

Because Fe is not translocated within the plant, deficiency symptoms first appear on the younger leaves at the top of the plant. Severe deficiency may turn the plant yellow-to-bleached white and can result in the death of plant limbs.

Iron deficiency may be caused by an imbalance with other metals such as Mo, Cu, or Mn. Other factors that may trigger Fe deficiency include:

- Excessive P in the soil,
- A combination of high pH, high lime, wet, cold soils, and high bicarbonate levels,
- Plant genetic differences, and
- Low soil organic matter levels.

Table 7-9 lists common Fe fertilizer sources and their Fe contents. Soil application or foliar spray can correct crop deficiencies to some degree. Applying soluble materials (such as ferrous sulfate; $FeSO_4$) to the soil is not very efficient because the Fe converts rapidly to unavailable forms. When these materials are applied as foliar sprays, they are much more effective. Injections of dry Fe salts directly into trunks and limbs have controlled Fe chlorosis in fruit trees. Most Fe fertilizer sources are best applied as foliar sprays. This method uses lower rates than soil application.

Altering soil pH in a narrow band in the root zone can correct Fe deficiencies. Oxidizable sulfur (S) products will lower soil pH and convert insoluble soil Fe to a form the plant can use. With some crops, such as soybeans, variety selection is the most effective means of managing Fe deficiency.

Table 7-9. Iron fertilizer sources and their Fe contents.

Source	Fe content, %
Iron sulfates	19-23
Iron oxides	69-73
Iron ammonium sulfate	14
Iron ammonium polyphosphate	22
Iron chelates	5-14
Other organics	5-10

Manganese

Manganese is taken up by plants as the divalent cation Mn^{2+}. It functions primarily as a part of enzyme systems in plants. It activates several important metabolic reactions and plays a direct role in photosynthesis by aiding chlorophyll synthesis. Manganese accelerates germination and maturity, while increasing the availability of P and calcium (Ca).

Because Mn is not translocated in the plant, deficiency symptoms appear first on younger leaves, with yellowing between the veins (see **Appendix A**). Sometimes a series of brownish-black specks appears. In small grains, grayish areas occur near the base of younger leaves. Manganese deficiencies are most common on high organic matter soils and on those soils with neutral-to-alkaline pH and naturally low in Mn content.

Although deficiencies are often associated with high soil pH, they may result from an imbalance with other nutrients such as Ca, magnesium (Mg), and Fe. Soil moisture also affects Mn availability. Deficiency symptoms are most severe on high organic matter soils during cool spring months when soils are waterlogged. Symptoms disappear as soils dry and temperatures warm. This condition could be the result of less microbiological activity in cool, wet soils. However, pH of these soils is also higher during the winter, and there is a close relation between extractable (available) Mn and pH.

Manganese deficiencies can be corrected in several ways:

- If liming caused the deficiency, keep soil pH below 6.5. This can be done by reducing aglime rates or using materials which produce soil acidification, including elemental S. Band applications, near but not in contact with seed, reduce pH and convert Mn into plant-available form. For field crops, it is more economical to add Mn than to try to lower soil pH.

- Mix soluble Mn salts, such as Mn sulfate ($MnSO_4$), with starter fertilizer and apply in a band. High P starter fertilizer helps mobilize Mn into the plant. A field deficiency symptom can be corrected by foliar application. Spraying with 10 lb/A $MnSO_4$ in water is a common treatment for Mn-deficient soybeans. **Table 7-10** shows common Mn fertilizer sources and their Mn contents.

Table 7-10. Manganese fertilizer sources and their Mn contents.

Source	Mn content, %
Manganese sulfates	26-28
Manganese oxides	41-68
Manganese chelate	12
Manganese carbonate	31
Manganese chloride	17

On some soils, an extremely acid pH may cause Mn toxicity to crops. Soil pH must be 5.0 or lower before significant toxicity threatens. Yet, toxic Mn levels in plants have been measured up to a pH of 5.8. Liming will eliminate this problem.

Molybdenum

Molybdenum is taken up by plants as the MoO_4^{2-} anion. It is required for the synthesis and activity of the enzyme nitrate reductase. This enzyme system reduces NO_3^- to NH_4^+ in the plant. Molybdenum is vital for the process of symbiotic N fixation by Rhizobia bacteria in legume root nodules. It is also needed to convert inorganic P to organic forms in the plant.

Molybdenum deficiency symptoms show up as a general yellowing and stunting of the plant (see **Appendix A**). A deficiency can cause N deficiency symptoms in legume crops such as soybeans and alfalfa because soil bacteria growing symbiotically in legume root nodules must have Mo to help fix N from the air.

A deficiency of Mo can also cause marginal scorching and cupping or rolling of leaves. Molybdenum becomes more available as soil pH goes up, the opposite of most other micronutrients. So, deficiencies are more likely to occur on acid soils. Sandy soils are deficient more often than finer-textured soils. **Table 7-11** shows the effects of Mo on soybean yields at varying soil pHs.

Table 7-11. Soybean response to Mo at varying pHs.

Soil pH	Yield, bu/A	
	Mo	Without Mo
5.6	41	32
5.7	43	34
6.0	40	35
6.2	42	40
6.4	41	42

Tennessee

Since Mo becomes more available with increasing pH, liming will correct a deficiency if the soil contains enough of the nutrient. This is illustrated in **Table 7-11**. Heavy P applications increase Mo uptake by plants, while heavy S applications decrease Mo uptake. Applying heavy amounts of S-containing fertilizer on soils with a borderline Mo level may induce Mo deficiency.

Several materials supply Mo and can be mixed with NPK fertilizers, applied as foliar sprays or used as a seed treatment. Seed treatment is probably the most common way of correcting Mo deficiency because of the very small amounts of the nutrient that are required.

Excessive Mo is toxic, especially to grazing animals. Cattle eating forage with excessive Mo content may develop severe diarrhea. Also, animals grazing pastures low in Mo may develop Cu toxicity if Cu soil levels are high enough. But, animals eating forage high in Mo may develop Cu deficiency, leading to a disease called molybdenosis. It can be corrected by feeding Cu sulfate ($CuSO_4$) orally, injecting Cu-containing medicines, or applying $CuSO_4$ to the soil. **Table 7-12** lists common Mo fertilizer sources, their Mo contents, and water solubility.

Table 7-12. Molybdenum fertilizer sources, Mo contents, and water solubility.

Source	Mo content, %	Water soluble
Ammonium molybdate	54	Yes
Sodium molybdate	39-41	Yes
Molybdic acid	47.5	Slight

Nickel

Nickel was added to the list of essential plant nutrients late in the 20th century. No deficiencies have been observed under crop growing conditions, but they have been produced in research and include:

- Chlorosis of young leaves.

- Death of meristematic tissue.

Nickel deficiency has been observed in some nursery plants and tree crops. Affected trees develop mouse-ear, a condition marked by small, curled leaves and stunted growth.

Nickel is absorbed by plants as the divalent cation Ni^{2+}. It is required in very small amounts, with the critical level appearing to be about 0.1 ppm. It is a component of the urease enzyme and is, therefore, necessary for the conversion of urea to ammonia (NH_3) in plant tissue, making it important in plant N metabolism.

Zinc is taken up by plants as the divalent Zn^{2+} cation. It was one of the first micronutrients recognized as being essential for plants and the one most commonly limiting yields. Although it is required in small amounts, high yields are impossible without it. Some crops are more responsive to Zn than others, as shown in **Table 7-13**.

Table 7-13. Responsiveness of crops to Zn.

Most response	Medium response	Least response
Beans	Barley	Asparagus
Corn	Potatoes	Carrots
Onions	Soybeans	Grass
Sorghum	Sudan grass	Oats
Sweet corn	Sugarbeets	Peas
Citrus	Table beets	Rye
Rice	Tomatoes	Cabbage
Peaches	Alfalfa	Celery
Pecans	Clover	Lettuce
Flax	Cotton	Grapes

Zinc aids synthesis of plant growth substances and enzyme systems and is essential for promoting certain metabolic reactions. It is necessary for production of chlorophyll and carbohydrates. Zinc is not translocated within the plant, so symptoms first appear on the younger leaves and other plant parts (see **Appendix A**).

Zinc deficiency in corn is called white bud because the young bud turns white or light yellow in early growth. Leaves may develop broad yellow bands (chlorosis) on one or both sides of the center mid-rib. Other symptoms include bronzing of rice, rosette of pecans, little leaf of fruit trees, and severe stunting of corn and beans.

Soils can contain from a few to several hundred pounds of Zn per acre. Fine-textured soils usually contain more Zn than sandy soils, but the soil's total Zn content does not indicate how much is available. Several factors determine Zn availability:

- **Soil pH**—Zinc becomes less available as soil pH increases. Some soils limed above pH 6.0 can develop Zn deficiency, especially sandy soils. Deficiencies do not occur on all soils with near-neutral to alkaline pHs. The tendency is simply greater. Zinc concentration in the soil can decrease 30-fold for every pH unit increase between 5.0 and 7.0.

- **High soil P**—Zinc deficiency may occur on soils with high P availability. Several plant species have shown Zn-P interactions. High levels of either Zn or P may reduce plant uptake of the other. Applying one on soils marginal in both may induce a deficiency of the other. Soil pH further complicates Zn-P interactions.

  Applying P to a soil with sufficient Zn levels will not produce a Zn deficiency. However, some consultants and laboratories caution that when P soil tests are high and annual P applications are still needed for high yields, one pound of Zn should be applied for every 20 lb of phosphate (P_2O_5).

- **Soil organic matter**—Much Zn may be fixed in the organic fraction of high organic matter soils. It may also be temporarily immobilized in the bodies of soil microorganisms, especially when animal manures are added to the soil. At the opposite extreme, much of a mineral soil's available Zn is associated with organic matter. Low organic matter levels in mineral soils are frequently indicative of low Zn availability.

- **Irrigation**—When soils are shaped or leveled for irrigation, Zn often becomes deficient because of organic matter removal, compaction, and exposure of high pH soil layers.

- **Leaching**—Zinc is adsorbed by soil colloids. This helps it resist leaching and stay in the topsoil.

- **Cold, wet soils**—Zinc deficiencies tend to occur early in the growing season when soils are cold and wet. This is due to slow root growth. The slow growing root system is unable to take up enough Zn to supply the upper portions of the plant. Plants sometimes appear to outgrow this deficiency, but the damage has already been done, and yields can still be significantly reduced.

- **Soil biological activity**—Zinc availability is affected by the presence of certain soil fungi, called mycorrhizae, which form symbiotic relationships with plant roots. Removal of surface soil in land leveling may remove the beneficial fungi and limit the plant's ability to absorb Zn.

The best procedure for correcting Zn deficiencies in field crops is to apply a Zn fertilizer broadcast preplant or in a starter fertilizer beside the row of planting. Application rates range from 1 lb/A of Zn to as high as 10 lb/A, depending on soil test levels. Very low rates should be applied in a starter. Zinc has excellent residual effects, and high application rates may be sufficient for several years. Check availability by soil testing. Zinc fertilizer sources are listed in **Table 7-14**.

Table 7-14. Zinc fertilizer sources and their Zn contents.

Source	Zn content, %
Zinc sulfates (hydrated)	23-36
Zinc oxide	78
Basic Zn sulfate	55
Zinc-ammonia complexes	10
Zinc chelates	9-14
Other organics	5-10

When soil tie-up of Zn is expected under high pH conditions or where an emergency situation exists on an established crop, Zn may be applied as a foliar spray. Foliar sprays usually require about 0.5 to 1.0 lb/A Zn. **Table 7-15** shows the effects of application methods of Zn on corn yields.

Table 7-15. Percentage of field trials and corn yield responses from Zn application methods.

Application method (no. of sites)	Percent of trials showing response of:			
	Less than 5 bu/A	5-10 bu/A	11-15 bu/A	More than 15 bu/A
Broadcast (31)	55	6	19	20
Row (28)	29	21	29	21
Foliar (31)	26	36	6	32

Kentucky

Zinc responses can be spectacular, as shown in **Table 7-16**. In this irrigated soybean study, a 4 lb/A Zn rate was the best treatment, producing a 20 bu/A yield response.

Table 7-16. Irrigated soybean responses to preplant Zn.

Zn rate, lb/A	Yield, bu/A	Leaf composition	
		P, %	ppm Zn
0	30	0.260	18
2	46	0.165	25
4	50	0.177	29

Kansas

Cobalt

Cobalt (Co) has not been proven essential for higher plant growth, but nodulating bacteria need it for fixing atmospheric N in legumes. SFM

Soil Sampling

Taking a good soil sample is the first step in a successful nutrient management program. It is important to understand and follow proper procedures. Sampling intensity should usually be based on expected variability in the field. Various approaches to sampling pattern can be selected, depending on conditions. Tillage system is also an important consideration in sampling procedure.

Development of a sound fertilizer and liming program must include soil testing, and soil testing must be based on samples that are representative of the area being tested. Otherwise, it will be of no value. Also, the entire process—sampling, laboratory analysis, interpretation, and recommendation—should be calibrated to and based on response data generated by field research.

The greatest potential for error in soil testing is in taking the sample. The one pound or less of soil submitted to the laboratory for testing represents anywhere from a few million to several million pounds in the field. The top 6 to 7 in. of soil—sometimes referred to as the plow layer or acre furrow slice—weighs about 2 million pounds. If one sample is taken for each 20 acres, it represents about 40 million pounds of soil. For a 2.5 acre grid, commonly used in site-specific nutrient management, the sample represents about 5 million pounds.

Actually, each analysis performed in the laboratory may require a teaspoonful of soil or less. Whether a pound of sample or a scoop of soil is used in the analysis, care in taking a representative sample is critical to the value of the end product—the soil test result used to make recommendations for fertilizer and lime and to develop whole-farm nutrient management plans.

Taking a representative soil sample is no easy task. The soil is a dynamic biological system. Variability results from many factors, including time of sampling, depth of sampling, moisture conditions, cropping system, fertilizer/lime/manure history, soil types, topography, etc. There is potential for significant variability in every field.

Review Questions:

1. (T or F) The greatest potential for error in soil testing is in taking the sample.

2. If a soil sample is pulled from 6 to 7 in. deep for each 20 acres in a field, each sample represents about _______ _______ pounds of soil.

Soil Sampling Procedures

What are the objectives of soil testing? One is to estimate the nutrient status of the soil; that is, the capacity of the soil to supply those nutrients necessary to meet the needs of the crop(s) to be grown. An additional objective is to identify variability in nutrient status of the area being sampled so that appropriate recommendations for fertilizer and lime can be made to provide, at non-limiting rates and amounts, sufficient nutrients to support the crops throughout the growing season. For field-average management, only one sample might be used in some situations, while multiple samples would be needed in the case of site-specific management. There are certain procedures that should be followed in obtaining representative samples, whether for field-average or site-specific nutrient management.

- **Be prepared.** Needed equipment and supplies include an auger or sampling probe, a clean plastic bucket, soil sample bags, a pencil or waterproof marking pen, and a field or farm map to identify the area being sampled—including a number and a sketch showing pertinent characteristics. (Note: Metal or rubber buckets contain minerals that may contaminate the sample, and thus should not be used.)

- **Divide fields into uniform areas.** Group by crop history, soil type, topography, soil color, degree of erosion, and any other obvious factors that might influence soil fertility status. If a person not familiar with the fields is taking the samples, it helps for someone to point out areas where homesteads or livestock areas existed in the past.

- **Sample at a uniform depth.** Nutrient concentration can vary significantly with soil depth, so care must be taken to sample at the same depth, in the collection of cores that make up a composite sample. For samples that will be taken in subsequent years, it is important to use the same sampling depth each time. **Figure 8-1** illustrates how variable nutrient content can be with soil depth.

- **Follow the specific instructions provided by the laboratory being used**. Review sampling depth, number

Figure 8-1. Effect of conventional tillage on P distribution [parts per million (ppm)] through two soil profiles. (CT = conventional tillage; ZT = zero tillage)

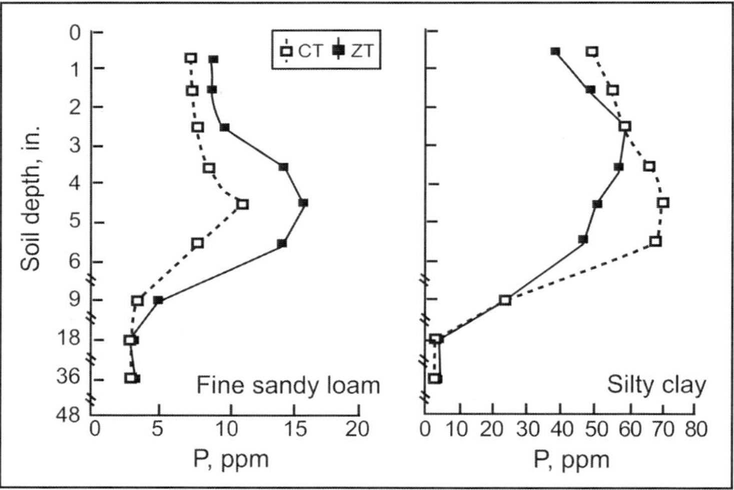

of cores per sample, number of samples per acre or field, sample drying procedures, field information requested, shipping instructions, etc. Most soil testing laboratories provide similar guidelines.

- **Use global positioning system (GPS) technology**, if available, to document the position of each sample. Recording the same location and other background data in a geographic information system (GIS) is the best way to prepare for interpretation of the results.

- **Make note on the field map of any special problem**

Soil Fertility Manual

areas that could be limiting yields, including weeds, diseases, wet spots, etc. Send this information to the laboratory, but also file it for future reference.

- **Be aware of the location of old fertilizer bands and crop rows**. These bands can have a dramatic impact on the validity of the soil sample. **Figure 8-2** is a hypothetical diagram that shows the influence of nitrogen (N) and phosphorus (P) fertilizer bands on nutrient variability. Over-sampling or under-sampling these bands can have a strong influence on the soil sample, and thus the resulting fertilizer and lime recommendation.

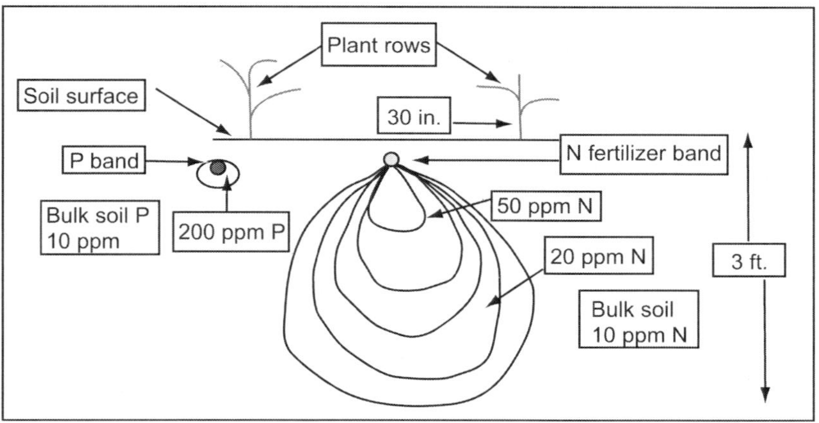

Figure 8-2. A hypothetical diagram shows the influence of N and P fertilizer bands on nutrient variability. The P band was placed 2 in. below and to the side of the seed and the N band was located in the center of the interrow area. This diagram was based on the findings of Stecker et al. (2000) and Clay et al. (1997). Source: SSMG #38, PPI/FAR

- **Sample at the proper time.** Samples should be taken to allow adequate time for test results to be returned from the laboratory so that recommendations can be prepared. It is important to sample at the same time each year if results are to be compared and soil fertility trends monitored. Taking samples for spring planted crops directly after fall harvest is ideal. Sampling during this period tends to minimize field variability. Mid summer samples should be avoided, especially on sandy soils, because of leaching. Samples should not be taken in late winter on heavy textured soils. Freezing and thawing tend to release potassium (K) and give unusually high soil test values.

Production Concept 8-1 provides general instructions for soil sampling.

General Instructions for Soil Sampling

Check with the laboratory that will handle your soil samples for analysis. They probably have more specific instructions appropriate for your area, crops, and management practices. The following are some basics.

A. You should have a sampling tube or auger and a plastic bucket. The lab where soil will be sent for analysis should have containers or bags for the samples.

B. Where possible, identify areas of the field that were treated differently in the past, and sample different soils separately. Take equal-sized cores or slices from 15 or more places in each sampling area using the probe or auger. Do not mix light- and dark-colored soils together.

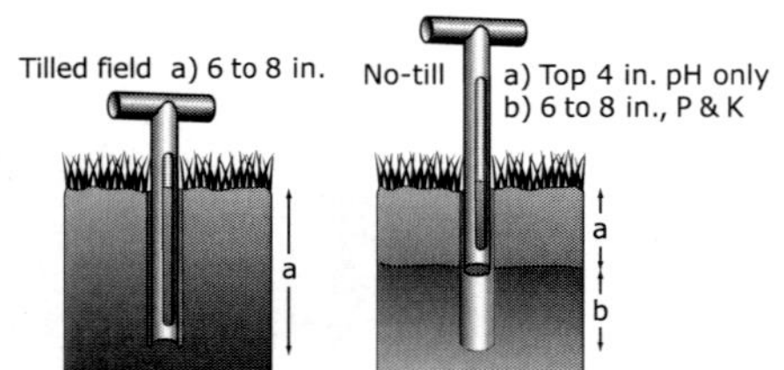

C. Generally, lab instructions call for samples for P, K, and aglime from the top 6 or 8 in. for no-till fields and forage stands, and from the top 0 to 4 in. for aglime only in no-till fields. No-till fields that will be plowed periodically should be sampled to plow depth.

D. Place cores or slices in the clean plastic pail. Mix them thoroughly, breaking up cores or slices. If soil is muddy, dry it before mixing. If soil crumbles easily, dry after mixing.

E. To dry, spread the mixture on a clean paper. Do not heat in an oven or on a stove. Do not dry where fertilizer or manure or other material may get in the sample.

F. Fill the sample bag with the air-dry soil to the level indicated by the lab. Discard the rest. Label and number the sample container.

G. Record the cropping and fertilizer information requested, using a field and cropping information sheet provided by the lab doing the analysis.

H. Keep a field map with current information in your records. It will be valuable for developing management plans and for other uses.

The number of samples needed—sampling intensity—should be based on the expected variability within the field. As an example, for a typical Midwest corn-soybean system, agronomists might recommend one sample per 2.5 acres or less. More intense sampling would be recommended if soil variability is expected to be high or where site-specific, variable-rate fertilizer applications are to be made. An intensity of one sample per acre might be preferred, especially in areas of high rainfall or where irrigation is used.

Obviously, both cost and time requirements for soil testing are higher as sampling intensity increases. These factors, along with other considerations, should be evaluated in making the final decision as to the number of samples to take. In the final analysis, the number should be sufficient to adequately represent the area being sampled.

Being familiar with a particular field helps in determining the intensity of sampling that might be necessary to characterize variability. If there is a wet spot or an area that once was a feedlot or homestead, the level of variability would be expected to be higher than if field history indicated more uniformity in cropping history and fertilizer application. Such a scenario might require that the field be divided into several sub areas, with intensive sampling being done in each area. The key is not to ignore known sources of variability, regardless of the sampling scheme being used.

Collecting only one sample made up of several cores or plugs results in a soil test that represents the average of the sampled area. Portions of the area will be under fertilized—areas where nutrient levels are below field average. Other portions will be over fertilized—areas where nutrient levels are above field average. To minimize under and over fertilized areas, fields should be split into sub fields on a uniform grid basis or divided by some other characteristic such as soil type, yield variability, etc., to create management zones. The size of the grids or management zones needed will depend on the amount of variability in the field and the value of the crop. Midwest farmers commonly use a 2.5 acre grid. Larger areas are more commonly used in the western Corn Belt and in small grain and forage areas. The result will be that more intensive sampling will be required than that necessary for field average representation. Other data layers that can help in determining sampling intensity and patterns include soil survey maps, topographic maps, yield monitor maps, remote sensing imagery, and aerial photographs.

Soil survey maps are useful in determining major limiting yield factors such as poor drainage, steep slopes, and erosion. Soil survey data can also be used to identify variation in soil organic matter, soil texture, and other factors influencing changes in moisture content across the field and over time. This is important information that can be used to help guide nutrient applications, pesticide rates, and other production inputs.

Increasing sampling intensity can help to identify areas in a field that should be treated differently than field-average

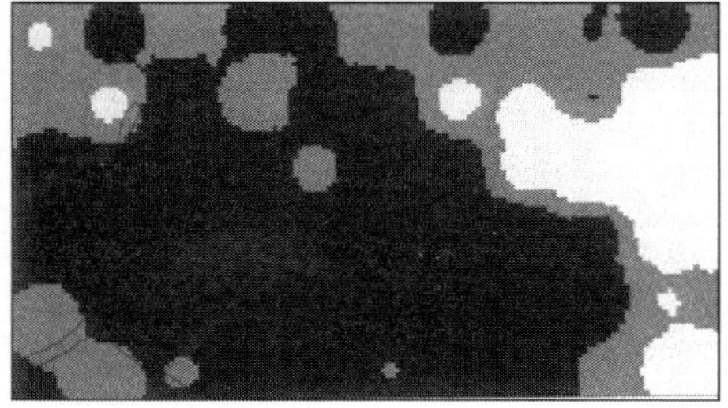

Figure 8-3. Fertilizer recommendations for 200 bu/A corn and 60 bu/A soybeans on a 90-acre central Illinois field.

Review Questions:

5. If soil variability is expected to be high where site-specific, variable-rate fertilizer application will be used, _______ _______soil sampling would be recommended.

6. Collecting one soil sample made up of several cores or plugs results in a soil test that represents _______ _______.

fertilizer recommendations. For example, a 90-acre central Illinois field was sampled on a 1-acre grid. The yield goals for a corn-soybean rotation were 200 bu/A for corn and 60 bu/A for soybeans. Following several years of field-average recommendations, the soil test K level (358 lb K/A) would require maintenance application only, according to the *University of Illinois Agronomy Handbook*. Based on the corn and soybean yield goals, the maintenance recommendation would be 134 lb K_2O/A for the two-year rotation.

Figure 8-3 shows the spatial distribution of soil test results based on the 1-acre grid sampling density.

When the results were interpreted, 47 acres showed a need for K build-up and maintenance applications, 30 acres needed maintenance only, and only 13 acres needed no K. Thus, the field-average interpretation correctly predicted K needs on only 30 of the 90 acres in the field and resulted in underfertilizing 47 acres and applying unneeded K fertilizer on 13 acres. In other words, two-thirds of the field was not properly assessed by the field average approach.

The field described above is representative of much of the eastern U.S. Midwest, where long-term fertilizer use has resulted in field-average soil test K levels being built to adequate ranges. However, significant areas in those fields still need build up applications to maintain optimum crop yields. At the same time, there are also areas that will require less than field-average recommendations.

Soil sampling with a modern hydraulic system on a pick-up.

Sampling for Conventional Management

Sampling the soil in an organized pattern that best represents the area being tested is a good management practice. This is true whether field-average or variable-rate fertilizer and lime applications are planned. For those fields that will be managed as a uniform unit, samples can be selected at random, but with the person doing the sampling being careful to cover the entire field–perhaps in a zigzag pattern or following a uniform grid pattern, as shown in **Figure 8-4**.

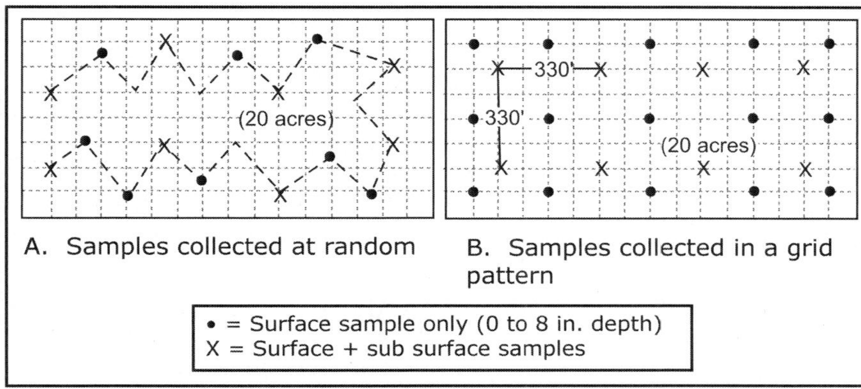

Figure 8-4. These illustrations compare patterns for soil samples collected at random to samples collected in a grid pattern.

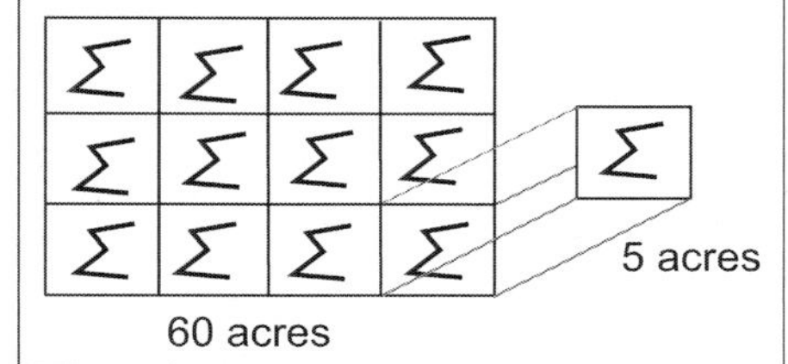

Figure 8-5. In this example for a 60 acre field, areas or "cells" of five acres each are identified, and five cores from each cell are collected in a zigzag pattern.

Use of a specific sampling pattern might be necessary to account for all the known sources of soil variability, such as past cropping patterns, major soil type changes, topography, fertilization history – including considerations for band-placed nutrients, etc. A grid pattern is usually the best way to be sure that the entire area being sampled is represented, but because of expected variability, it is advisable to utilize a scheme that avoids arranging sampling points in a straight line.

In the Midwest, a common approach is to divide the field into cells ranging in size from 2.5 to 5 acres each and collecting five cores in a zigzag pattern throughout the cell for each sample, as shown in **Figure 8-5.** This method allows for fairly complete sampling of the field, with good representation of the nutrient requirements for a single uniform application of the entire area to be fertilized.

Sampling for Site-Specific Management

To better characterize a field for site-specific management and variable-rate application, **point** samples within each cell should be used to measure the variability across the field. If a field is divided with a 2.5-acre grid and each cell is sampled, the grid lines help to ensure a good spatial representation of the field that can be used to develop a soil fertility (nutrient) map. At least five cores should be collected for each sample, and they should be within a 10 ft. radius of the center point of the sample. This procedure provides the basis for nutrient variability maps. With more points (smaller grids) the sample will be more representative of the field. Many have concluded that

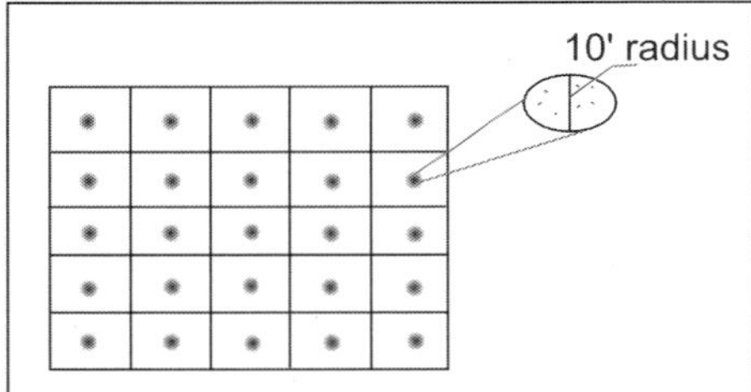

Figure 8-6. With the grid point sampling technique, soil test values represent a point (stratified systematic square grid).

the practical and economic limits to sample density is about 2.5 acres per sample. See **Figure 8-6.** In the western Corn Belt, it is more common to use a larger number of cores per sample, with each sample sometimes representing a larger area. The amount of variability, local research, and economics influence the decisions on grid size and number of cores per sample. Experience and local research are important considerations.

To avoid sampling bias caused by patterns in the field due to tillage, crop residue, fertilizer application, etc. associated with crop production, a staggered sample pattern, as shown in **Figure 8-7**, can be used.

This pattern helps to avoid the pattern bias, yet provides an organized sampling scheme to represent the entire field. It can be set up by counting rows, using a measuring wheel, or using GPS navigation. To gain the benefits of grid sampling and retain the benefits of random sampling, the stratified systematic unaligned sampling pattern, as illustrated in **Figure 8-8** can be useful in helping to avoid the effects of any patterns in the field.

Use of GPS technology provides for accurate positioning of sample points, allowing production of accurate geo-referenced

Figure 8-7. (Left) The staggered sampling pattern can be a triangle, diamond, or hexagon.

Figure 8-8. (Right) The stratified systematic unaligned sampling approach avoids any patterns in the field.

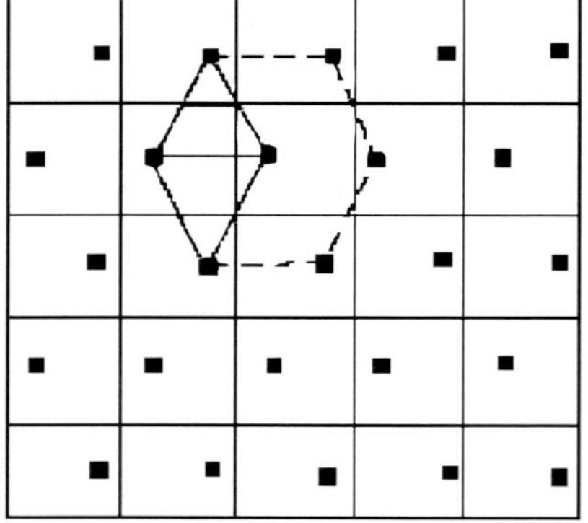

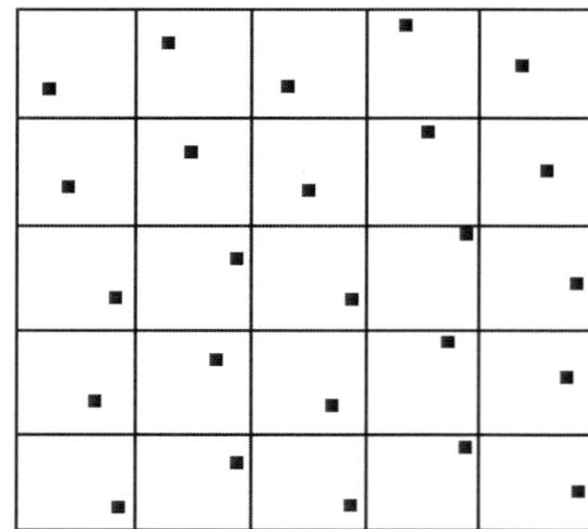

maps of nutrient levels with GIS and related to other data sets such as yield maps, soil surveys, and remote sensing imagery. Even if GPS is unavailable, location of sample points should be referenced by some means, so that approximate maps of variability can be drawn.

Sampling by Soil Type

Sampling by soil type requires a good soil survey map for the field. Such maps are available from the Natural Resources Conservation Service (NRCS) or soil survey units. As digital soil surveys are completed for many counties in the U.S. and Canada, they can be incorporated into the GIS database, making all the data associated with soil type available for use in developing fertilizer and aglime recommendations and nutrient management programs.

When sampling by soil type, sample points should lie within the boundaries of the soil type being sampled. Care should be taken to avoid taking samples in the transition areas between two soil types, to allow the full expression of the influence of soil type on the interpretation of soil test results.

As with grid sampling, it will be necessary to choose between area sampling—several cores taken as random points throughout the soil type boundary and mixed together for the sample—or point sampling, where several cores are collected within a few feet of specific sample points within the boundaries of each soil type. If point sampling is used, the points should be geo-referenced if possible so that they can be related to other data sets or to future soil sampling.

The number of samples per field should be based on known variability within the field. The number of cores per sample can also be determined on the basis of variability.

Sampling by Management Zone

After all is said and done, soil samples should be taken to represent the area being tested and to reflect the management plan of the grower. The plan could simply be to take one composite sample per field and fertilize accordingly. This approach to soil testing might work on small, uniform fields, but it does little to identify variability within a field.

At the other extreme is sampling at a very high intensity, perhaps on a 1-acre grid or less. Such an approach can certainly identify much of the variability in a field, but it is not practical for many growers. Sampling at such a high intensity is expensive and time consuming. Not many growers are willing to spend that much time and money on their nutrient management programs.

Most farmers should be managing their fertilizer and aglime program at a level that falls somewhere between the two discussed above. Their soil testing programs should be designed to fit their overall management skills and level and, just as important, assure that the environment is protected in the process.

Production Concept 8-2 compares how three different approaches to soil testing might compare in a non-uniform field.

Soil Sampling—Conventional, Grid, and Management Zone

Not every farm field is perfectly square and flat. More often, there are variations in elevation, drainage, cropping history, and other factors which must be considered. These illustrations compare how three different approaches to soil sampling might compare in the same field.

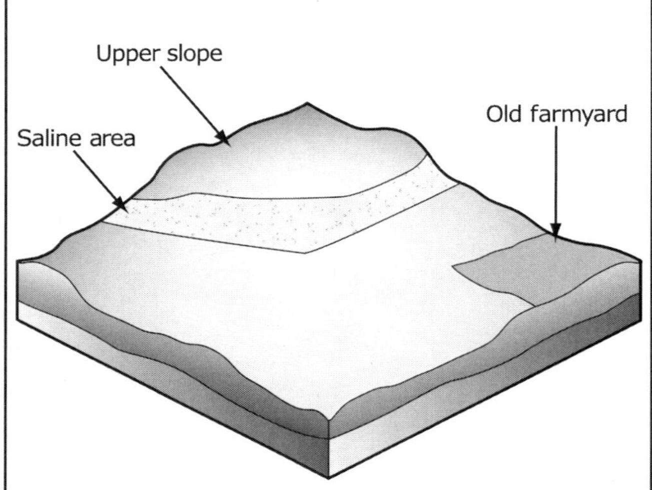

A. The field.

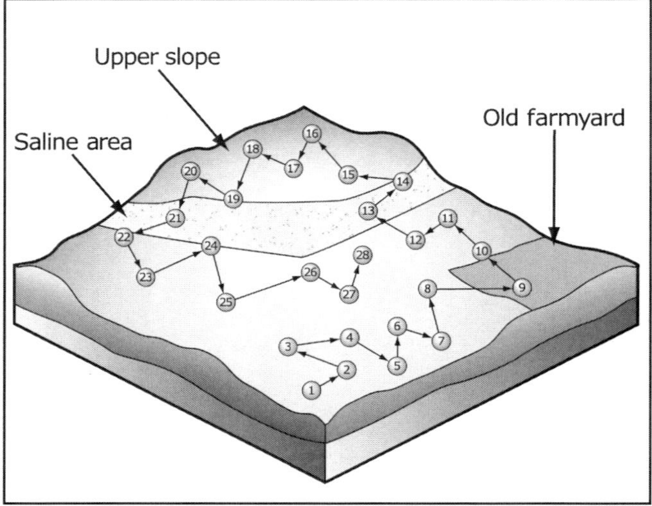

B. Conventional sampling.
Each ◯ actually represents a soil core, with a minimum of 15 to 20 cores making up the composite.

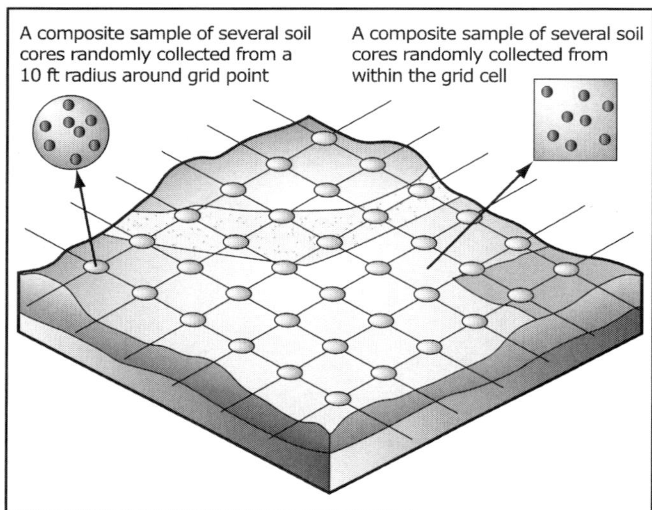

C. Grid sampling.
Sampling soil at equally spaced distances throughout the field either around a grid point or within a grid cell and analyzing each sample separately.

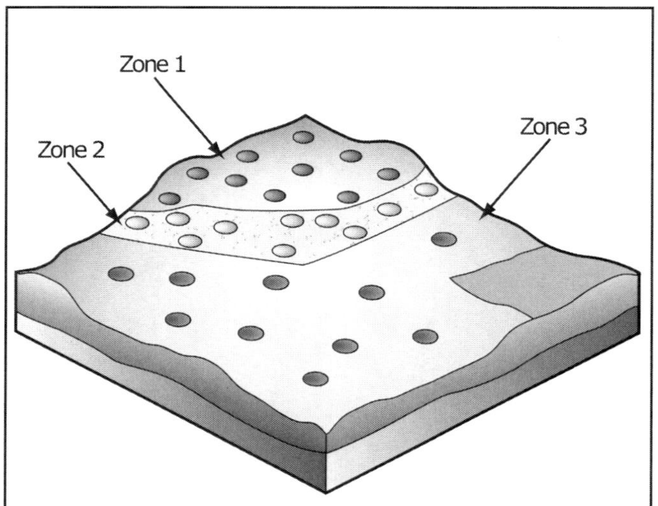

D. Sampling by management zone.
Taking a composite sample made up of 15 to 40 soil cores from areas of the field that can be managed separately.

The tillage system used will influence the degree of mixing of nutrients in the soil as well as nutrient movement through the soil profile. Often nutrients become stratified, or layered at various depths. This can affect nutrient availability of the growing crop, especially during periods of moisture stress that might be severe enough to limit root growth and activity during the growing season. For example, in a minimum tillage system, nutrients can accumulate in the top 3 to 4 in. of the root zone—an area that can dry out quickly during a hot, dry period in mid summer. As a result, crops can suffer nutrient shortages because of positional unavailability. Tillage systems being used should be a consideration when a field is sampled because of the influence tillage has on nutrient levels and movement in the soil profile.

If a moldboard plow is used at least every two or three years, nutrients such as P and K tend to be uniformly distributed through the plow layer. Soil pH values are usually fairly uniform as well. For P, K, and aglime recommendations, samples should be taken to the depth of plowing—usually about 6 to 8 in.

Some nutrient and pH stratification can be expected with mulch tillage—including chisel, disk, and field cultivator systems. Samples for routine soil analysis, including pH, P, and K, should be taken to a depth of about 6 to 8 in., being careful to avoid crop rows and fertilizer bands. Because mulch tillage helps to retain soil moisture, stratification is not usually a problem and, in fact, may result in concentrations of nutrients in small zones of varying pH that can enhance nutrient uptake efficiency.

Distinct stratification of soil pH and nutrients such as P and K occurs where continuous no-till is practiced. Samples for routine P and K analysis should be taken to a depth of about 6 to 8 in., but avoiding crop rows and fertilizer bands. Stratification under no-till is not usually a problem. However, long-term no-till fields can become nutrient deficient in the lower part of the old plow layer during periods of drought stress.

In no-till systems, monitoring the 4 to 8 in. depth, especially for K, might be helpful. Deep band placement of K is an effective way to overcome the effects of drought stress.

Aglime is relatively immobile in the soil. Thus, recommendations for continuous no-till fields, where aglime is surface applied, should be based on a 4 in. sampling depth. This also means that the aglime recommendation appropriate for a conventionally tilled field (sampled at an 8 in. depth) at the same pH should be cut in half.

Sampling for Home Gardens and Lawns

Soil testing is also a valuable tool for improving home gardens and lawns. **Production Concept 8-3** illustrates some suggestions.

Soil Sampling for Home Gardens and Lawns

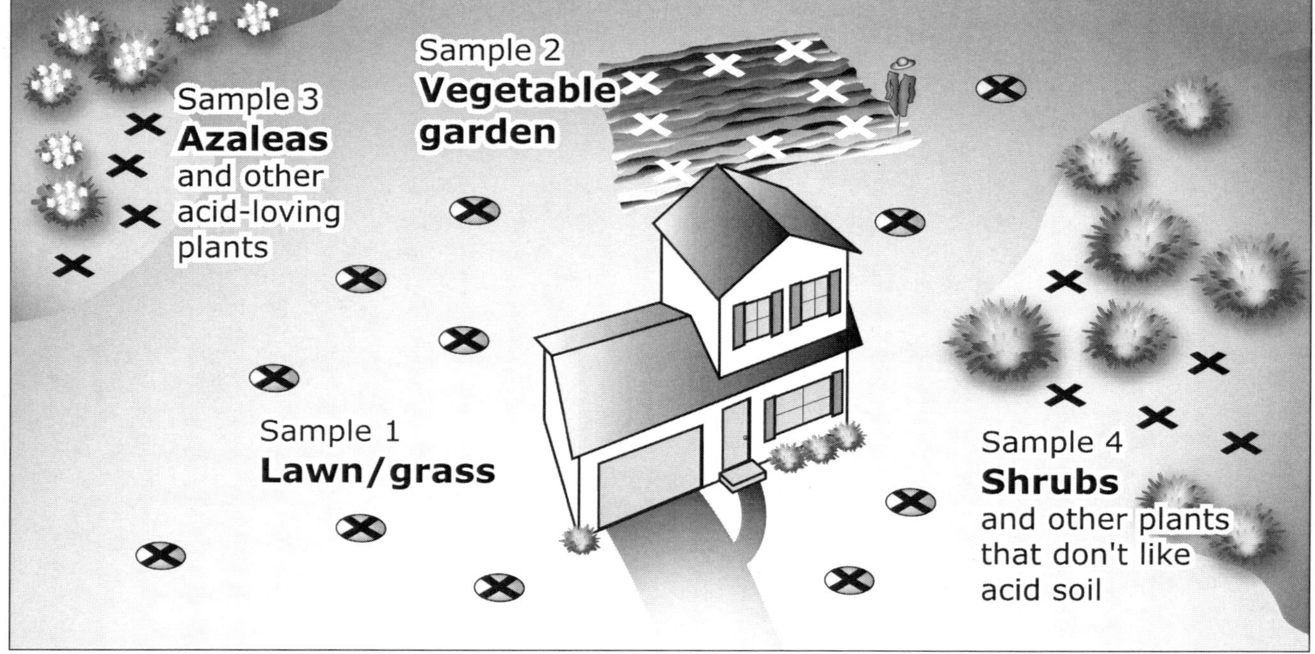

While soil testing can be a valuable practice for crop production, it can also be quite beneficial in typical home landscape settings. Although more specific instructions will be available from the local Extension office, garden center, or other service provider, here are some basics to remember. A clean plastic bucket and a shovel, trowel, or sampling probe are the first tools to find.

To begin, step back and consider the different soils and soil management situations you need to sample. If your only concern is to have a green, healthy lawn, then simply collect some random plugs (cores) of soil and combine them to send in one sample. But suppose you also have other zones that have different requirements for the plants you want to grow. For example, the illustration shown here identifies four distinctive growing areas or zones that would require separate samples to guide fertilizer recommendations.

Besides the soil sample for the lawn area, this property has a vegetable garden, an area for growing azaleas and other acid-loving plants, and yet another area for shrubs and other plants that don't do well in acid conditions. Generally, 10 to 15 plugs of soil, each about 4 in. deep (into the root zone) will be adequate to give a representative sample for each separate area. Proper soil pH range for the plants you are trying to grow is one of the most critical factors. Effectiveness of fertilizer depends largely on soil pH.

As you combine the plugs from one area into the bucket, mix thoroughly and remove any worms, stones, or plant matter. Scoop out about a pint of the sample, put it into a paper bag, and label its source. Repeat this process for each of the different zones you want to have tested. Carefully follow the guidelines of the soil testing laboratory.

Summary

Taking a good soil sample is the first step to a successful soil testing program. The sample—usually a pound or less—represents millions of pounds in the field. It had better be the right one. There are several keys to successful soil sampling, including—but not limited to—the following:

- Collect a sufficient number of samples to adequately represent the area being tested.

- Follow instructions provided by the laboratory chosen to do the testing.

- Sample at the proper time. The period following harvest is an ideal time for spring planted crops such as corn and soybeans. Avoid midsummer sampling on sandy soils and late winter sampling on heavy textured soils.

- Sample at the same time, the same location, and the same depth with follow-up samples so comparisons can be made to the original sample. Otherwise, changes in the nutrient management program cannot be accurately monitored.

- Follow a sampling program that meets the management level and needs of the grower. ⟨SFM⟩

Chapter 9

Soil Testing, Plant Analysis, and Diagnostic Techniques

Soil testing is a valuable tool for modern production agriculture. The most benefit comes when soil test information is used with all other available information to help make recommendations for greater yields and profits. Plant analysis is usually considered as a tool to supplement soil testing, and data should be interpreted by a qualified individual. Quick field tissue tests can also be useful in some situations. While knowledge of deficiency symptoms is important, "hidden hunger" in crops can reduce yields and quality before visual signs appear. To properly diagnose fertility problems or other conditions, the entire crop environment…from root zone to tillage practices…should be considered.

Soil Testing

When soil testing was introduced during the first half of last century, it brought with it considerable controversy. Certain university administrators viewed it with disdain. A few even equated it to black magic and would not allow it to damage the reputation of the local Cooperative Extension Service by banning it from campus, so to speak. There have always been detractors, but soil testing has become an integral part of modern production agriculture.

During its early stages of development and after it was accepted and put to widespread use, the primary role of soil testing was to determine the agronomic needs for fertilizer and aglime in crop production. While basic fertilizer/aglime requirements are still important and will remain so, soil testing has taken on a much broader role in nutrient management in recent years, extending to environmental considerations. In particular, it is useful in evaluating potential environmental impacts of nitrogen (N) and phosphorus (P) fertilization. For more detail regarding N and P fertilization and the environment, see Chapters 3, 4, and 11.

An Essential Management Tool

Soil testing will remain an essential management tool, utilized by industry and university agronomists, consultants, and farm managers for the benefit of their farmer clients. Fertilizer dealers will continue to use it as a planning and marketing aid and as a management support service for their customers.

The environment benefits from the improved protection of soil and water resources when nutrient management plans are written for specific farmer fields and specific areas within fields. So do farmers. That is, site-specific farming—including nutrient management—can produce higher, more efficient yields while improving future production potential by protecting and enhancing the environment. Soil testing plays a key role in site-specific farming.

Soil testing is extremely valuable in helping to prepare individual fertilizer and aglime recommendations and in developing whole-farm nutrient management plans. Soil tests serve many functions.

- They identify yield-limiting factors, specifically nutrient shortages in the soil.

- They indicate the nutrient supplying capacity of the soil being tested and, hence, where to start in developing fertilizer and aglime recommendations.

- Soil test results, combined with other production information—such as cropping history, soil survey maps, yield maps, and other data—are valuable input as nutrient management plans are prepared.

- They can be used on a regular basis to monitor soil fertility levels and to measure trends and changes, so that nutrient management can be kept on track with other production inputs.

- They can help manage risk. Crops grown on low testing soils are more likely to respond to fertilization in the year of application, but they also run the risk of not yielding as much as crops grown on higher testing soils. Soil fertility levels can be determined and monitored only through soil testing on a regular basis.

Farmers and their advisers need to know what tests to request when submitting a soil sample to a laboratory for testing. Years ago, the most common tests were for P, potassium (K), and aglime (soil pH). Now, with ever increasing yields placing greater demands on the soil, the farmer needs to know as much about the soil nutrient status as is possible, including information on sulfur (S), magnesium (Mg), and the micronutrients. Such tests can be done with precision in most laboratories. Nitrate-N (NO_3–N) tests are effective in assessing N fertilizer requirements. They work particularly well in lower rainfall areas, but are also of value in more humid areas as well.

Taking Soil Samples

In soil testing, the greatest potential for error is in field sampling. Since a few ounces of soil analyzed in the laboratory represent millions of pounds in the field (each 6-in. furrow slice of soil weighs about 2 million pounds), collection of a representative sample is critical. If a good sample is taken, the results of the analysis will provide a reliable estimate of the nutrient status of the soil. For details on soil sampling, see Chapter 8.

Interpreting Soil Test Results

Soil testing—properly utilized—is an excellent guide in determining fertilizer/aglime recommendations and in the development of nutrient management plans. However, it is not an exact science. Considering the dynamic nature of farming, differences in management styles and practices, and other factors such as weather, variability from year to year—and even within individual fields in a given year—is to be expected. (See Chapter 8 for help in understanding soil variability and how to use different sampling techniques to remove as much of the variability as possible.) That is why it is critical that the person responsible for the interpretation of soil tests and the preparation of recommendations be well trained and experienced.

Farmers who have their soils tested do so because they are interested in increasing their crop yields and profits. They also want to build soil fertility and long-term productivity while protecting the environment. A sound recommendation should take economic, agronomic, and environmental factors into consideration, including the following:

- recommendations for optimum crop yields so that all nutrients will be maintained at non-limiting levels from crop planting to harvest;

- balanced recommendations to ensure that each nutrient is used in the most efficient manner by the growing crop;

- for soils which test medium or less, buildup plus maintenance nutrient rates to bring soil test levels into the optimum range are usually appropriate. A testing schedule that will monitor soil nutrient levels—assuring that they remain in the target range—should be followed;

- in cases where P buildup has occurred because of use of excessive amounts of manure, exclude P from the fertilizer recommendation for one or more years.

Those who interpret soil test results know that the probability of response to nutrients such as P and K is greater when soils test medium or lower. However, crops grown on soils testing sufficient often respond to nutrient applications, particularly starter fertilizers. Such responses are more common when farmers are managing for higher yields, including early planting in cool, wet soils. Conservation tillage practices, soil compaction, and extreme soil pHs (both high and low) also increase the chance of response to P and K, especially from starter applications.

Farmers looking for greatest profits should ask for fertilizer recommendations that will accommodate high yields, but they also need more than a fertilizer recommendation. They need a complete nutrient management plan in addition to information on proper varieties, cultural practices, time of planting, appropriate use of pesticides, etc. Remember, a soil test, properly analyzed and interpreted, is only one part of an overall management plan that will assure high, profitable, and efficiently produced yields and, at the same time, minimize impact on the environment.

Soil Fertility Considerations

Fertilizer use and crop yield statistics indicate that soil fertility on many farms in North America may be declining due to less than optimum nutrient management. Further, a 2001 soil testing survey conducted by the Potash & Phosphate Institute found that nearly half of the more than 2.5 million samples included tested medium or lower in P and/or K. The consequences of 'mining' soil nutrients might not become apparent for several years, but will include lower crop yields and greater potential damage to the environment.

When little or no fertilizer is applied to soils, nutrient levels can drop quite rapidly. For example, when no P and K are applied in a corn-soybean rotation, even with high or very high soil tests, the P soil test (Bray P-1) will typically drop 5 to 6 lb/A/yr, and the K test (ammonium acetate extractable) will decline approximately 10 to 15 lb/A/yr. If under-fertilized, many high-yielding forage crops could lower the P test 6 to 8 lb/A/yr and the K test as much as 80 lb/A/yr. If such trends were allowed to continue—as is now being observed in many areas of North America—substantial loss in crop yields will result. Soil and water resources could also be damaged. It might take several years of higher fertilizer rates to restore optimum productivity.

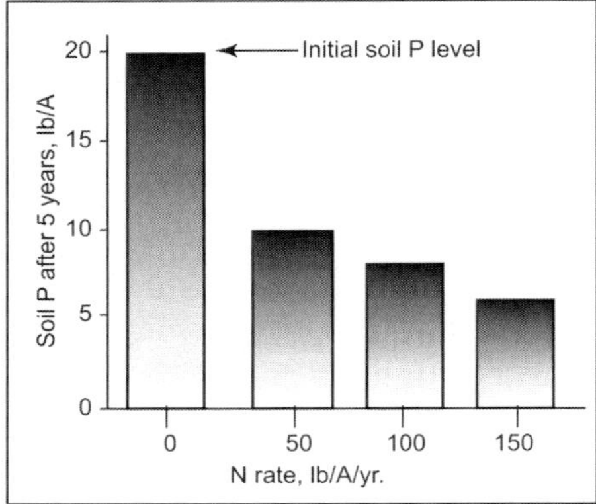

Figure 9-1. Effect of increasing N rates on soil P levels after five years of winter wheat (Texas).

It has been estimated that 6 to 14 lb of P_2O_5/A—above the P removed by the crop—would be required to raise the soil test P level by 1 lb. About 4 to 8 lb K_2O/A, also above crop removal, would be needed to raise the K soil test by 1 lb. The actual amounts, of course, would depend on several factors, including initial soil test levels, soil texture, clay minerals present, organic matter level, tillage system, etc. Farmers should consider a 4- to 8-year approach to raising soil test P and K to optimum levels. It is possible to do it more quickly, but the economics are usually less attractive.

Allowing soil tests to decline is destructive to future productivity, profitability, and the environment. Such declines can occur rapidly, as noted above. **Figure 9-1** shows how increasing N rates without adequate additions of fertilizer P drew down soil test P after only five years of winter wheat production.

Review Questions:

7. Fertilizer use and crop yield statistics indicate that soil fertility on many farms in North America may be _______ due to less than optimum nutrient management.

8. When no P and K are applied in a corn-soybean rotation, even with high or very high soil tests, the Bray P-1 soil test will typically drop _______ lb/A/yr and the K test (ammonium acetate extractable) will decline about _______ lb/A/yr.

Choosing a Soil Testing Laboratory

Care should be taken in selecting a soil testing laboratory. One should be chosen that offers the latest in analytical accuracy, can perform the analyses desired (and can do so in an appropriate period of time), and uses acceptable nutrient extraction methodology. Also, remember that the number generated during laboratory analysis does not indicate the absolute soil level of the nutrient being evaluated. Rather, such readings are simply indices of soil nutrient content and are of little value unless they can be correlated to crop response data that have been generated through years of research. Be sure the laboratory chosen uses research based analytical methodology.

Since one of the most important functions of soil testing is

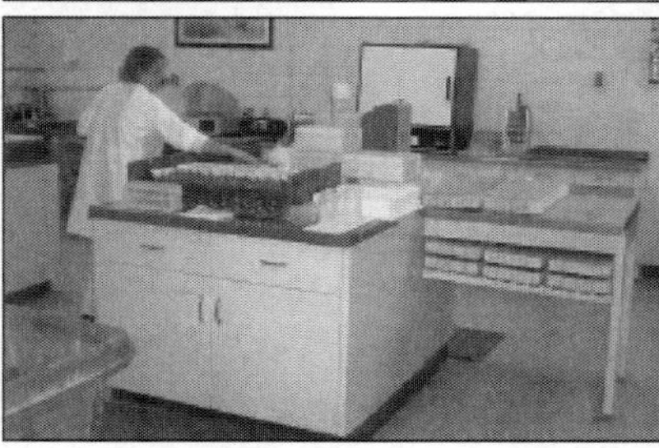
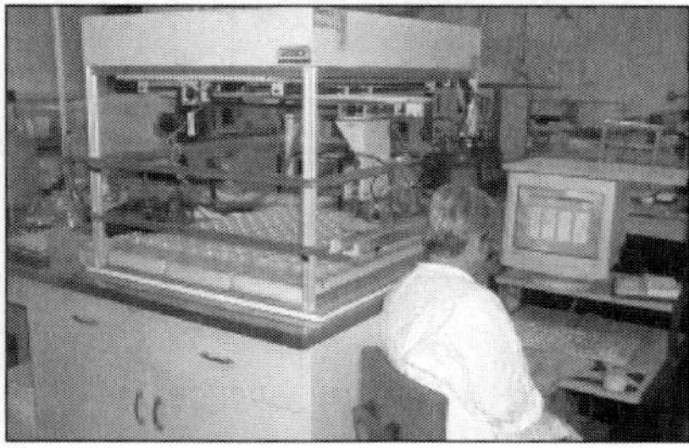

Modern soil testing laboratories offer a wide variety of tests and the latest in computer and analytical technology.
Photos courtesy of University of Arkansas Soil Testing Laboratory.

to monitor soil fertility trends over time, the laboratory being used must be able to provide accurate, reproducible results. Otherwise, agronomists and other advisers will not be able to determine if trends are real or the result of laboratory variability. The North American Proficiency Testing (NAPT) program, managed through the American Society of Agronomy, provides for national sample exchange for agricultural testing laboratories. Participation in the NAPT program should be a significant determining factor when a laboratory is being evaluated.

Once a laboratory has been chosen, continue with it. Switching from one laboratory to another can create confusion in interpreting results collected over a period of time, thus reducing the monitoring value of soil testing. Also, avoid splitting samples and sending to two or more laboratories. That practice usually leads to additional confusion and frustration. Besides, the NAPT program is already doing the splitting. Finally, changing testing laboratories should be done only if service, dependability, cost, precision, and other factors dictate the selection of another testing source.

Plant Analysis

The term 'plant analysis' refers to the total or quantitative analysis of nutrients in plant tissue. Plant analysis should not be confused with field tissue testing, a qualitative measure of nutrient content. (Tissue testing is discussed in a later section of this chapter.)

Soil testing and plant analysis are highly complementary and go hand in hand in evaluating soil fertility and overall nutrient availability. However, each has its strengths and limitations. Plant analysis assesses nutrient uptake while soil testing predicts nutrient availability. Soil testing is not always a good indicator of the availability of nutrients such as N and S—those nutrients that leach easily. Plant analysis is a better tool for assessing some micronutrients, including boron (B), iron (Fe), and molybdenum (Mo). On the other hand, plant analysis cannot predict aglime requirement.

In its early years, plant analysis was used primarily as a diagnostic tool in the correction of nutrient deficiencies during subsequent crop years. Now, though, it is commonly used to identify in-season nutrition problems in time to take corrective action on the current crop. It adds accuracy to the monitoring process as nutrient management plans are implemented.

Plant Nutrient Sufficiency Ranges

The value of plant analysis as a diagnostic tool is largely dependent on how results are interpreted. Usually, the procedure for interpretation is to compare elemental concentrations found in plant tissue with some standard sufficiency range for normal plants as established by research. **Table 9-1** shows typical ranges in nutrient concentrations in leaf tissues for various plant species. It should be noted that nutrient concentrations can vary with crop, variety, plant part sampled, growth stage when the sample is taken, environment, geographic area, and other factors.

Where research to establish plant nutrient sufficiency ranges has not been done, plant analysis can still be useful in identifying nutrient stress problems if paired plant samples can be taken from both 'poor' and 'good' growth areas in the field—where the same variety, soil moisture, and environmental conditions exist.

Table 9-1. Typical ranges in concentration of macro and micronutriets in mature leaf tissue for various plant species (Munson 1998).

Nutrient	Deficient	Sufficient or normal	Excessive or toxic
Macronutrients	- - - - - - - - - - - % - - - - - - - - - - - - -		
Nitrogen (N)	<2.50	2.50-4.50	>6.0
Phosphorus (P)	<0.15	0.20-0.75	>1.0
Potassium (K)	<1.00	1.50-5.50	>6.0
Sulfur (S)	<0.20	0.25-1.00	>3.0
Calcium (Ca)	<0.50	1.00-4.00	>5.0
Magnesium (Mg)	<0.20	0.25-1.00	>1.5
Micronutrients	- - - - - - - - - - - ppm - - - - - - - - - - -		
Boron (B)	5-30	10-200	50-200
Chloride (Cl)[1]	<100	100-500	500-1000
Copper (Cu)	2-5	5-30	20-100
Iron (Fe)	<50	100-500	>500
Manganese (Mn)	15-25	20-300	300-500
Molybdenum (Mo)	0.30-0.15	0.1-2.0	>100
Zinc (Zn)	10-20	27-100	100-400

[1] Recent field studies have redefined the Cl-deficient level for wheat as 1,000 ppm in a whole, above-ground plant at heading.

Plant Analysis Serves Many Roles

As with soil testing, plant analysis has advanced in its value as a diagnostic tool largely because of field research that can be correlated to laboratory analyses and the great strides made in analytical methodology. It serves many roles in advancing nutrient management. Plant analysis can be used to:

- confirm a diagnosis made from visible symptoms;
- identify 'hidden hunger' where no symptoms appear;
- locate soil areas where deficiencies of one or more nutrients occur, but which may not have been identified by soil testing;
- determine whether fertilizer nutrients have entered the plant;
- learn about interactions among various nutrients;

- when performed on the portion of the crop harvested, can provide site-specific nutrient removal estimates;

- evaluate the need for additional tests or studies in identifying a crop production system;

- help to evaluate extreme fluctuations in nutrient concentrations brought on by stresses other than nutrient deficiencies—such as drought, heat, cold, insects, diseases, etc.;

- identify potential toxic levels of nutrients...toxicities can occur when plant growth and yield potential decrease while nutrient concentration in the plant continues to increase (**Figure 9-2** shows the relationship between nutrient concentration in the plant and crop yield);

- help optimize yield, nutrient use efficiency, and environmental protection.

Figure 9-2. Relationship between nutrient concentration in the plant and crop yield (adapted from Havlin et al. 1999).

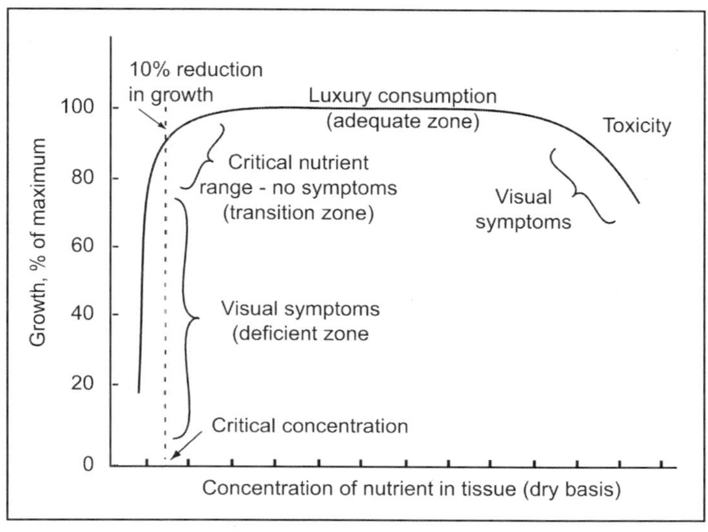

Follow Laboratory Instructions

An important step in a successful plant analysis program—just as is the case with soil testing—is sample collection. Plant composition varies with age, the portion of the plant sampled, the condition of the plant, the variety, weather, and other factors. Therefore, it is necessary to follow proper sampling procedures. Otherwise the accuracy of the laboratory analysis will be questionable, and proper interpretation of tests cannot be made.

Laboratories provide instructions for sampling various crops, plus information sheets and directions for handling and shipping samples. They usually suggest that samples from both good and problem areas of fields be included for comparison. They may also ask for companion soil tests representing the same areas. It is essential that instructions from the laboratory be carefully followed and that all information requested be provided to the laboratory–such as soil type, moisture conditions, landscape position, cropping history, fertilizer applications, use of pesticides, and a description of the problem or deficiency.

All nutrient management plans should be flexible and subject to periodic evaluation, including timely soil testing and plant analysis. Plant analysis helps ensure that soil fertility objectives are being met and that nutrition does not limit yield potential. Coupled with a sound soil testing program and other production management inputs, it can be an excellent diagnostic tool, for both in-season corrective treatments or more long-term adjustments in nutrient management plans.

Significant advancements have been made in plant analysis, in terms of research and analytical methodology. However, much remains to be learned about this diagnostic tool. Research is constantly uncovering new facts and establishing new standards. Data should be interpreted by personnel trained in the field and who understand the factors involved.

Remember: Yield losses due to a nutrient deficiency usually occur before the deficiency is severe enough to cause the crop to exhibit visual symptoms. It is important that soil testing and plant analysis be used to develop nutrient management programs so that soil fertility does not become a limiting factor to high yield, high profit crop production.

> **Review Question:**
> 17. (T or F) Yield loss due to nutrient deficiency should not be expected unless the crop shows visual symptoms of deficiency.

DRIS

DRIS (Diagnostic and Recommendation Integrated System) is a mathematical technique to apply plant analysis information for diagnosing the most limiting nutrient in a production system. The evaluation is made by comparing the relative balance of nutrient content with norms established for that crop under high yield conditions. Nutrient balance is a part of the proper interpretation of DRIS systems because nutrient interactions...to a large extent...determine crop yield and quality. **Production Concept 9-1** illustrates how nutrient interactions can affect yields.

There are certain limitations to the use of the DRIS system in production agriculture in North America. The system compares calculated nutrient ratio indices with established norms. The challenge with DRIS is in the development of the norms, which consist of thousands of data entries from a range of randomly selected high yielding populations. The norms must be based on local data to be applicable for interpretation of local plant analysis. In many cases, those data are not available. Relatively few laboratories use the DRIS method of plant analysis interpretation—partly because of the lack of field verification, calibration and correlation of data, and unfamiliarity with the system. Also, research comparing DRIS to nutrient sufficiency range values has shown that DRIS-based interpretations are often not superior.

> **Review Questions:**
> 18. To be an effective diagnostic tool, plant analysis should be used in conjunction with _____ _____.
> 19. (T or F) DRIS is a mathematical-based technique used to determine nutrient sufficiency range values for plant analysis interpretation.
> 20. The DRIS method of plant analysis interpretation is not routinely used in all laboratories because _____.

Nitrogen Increases Uptake of Other Nutrients by the Plant

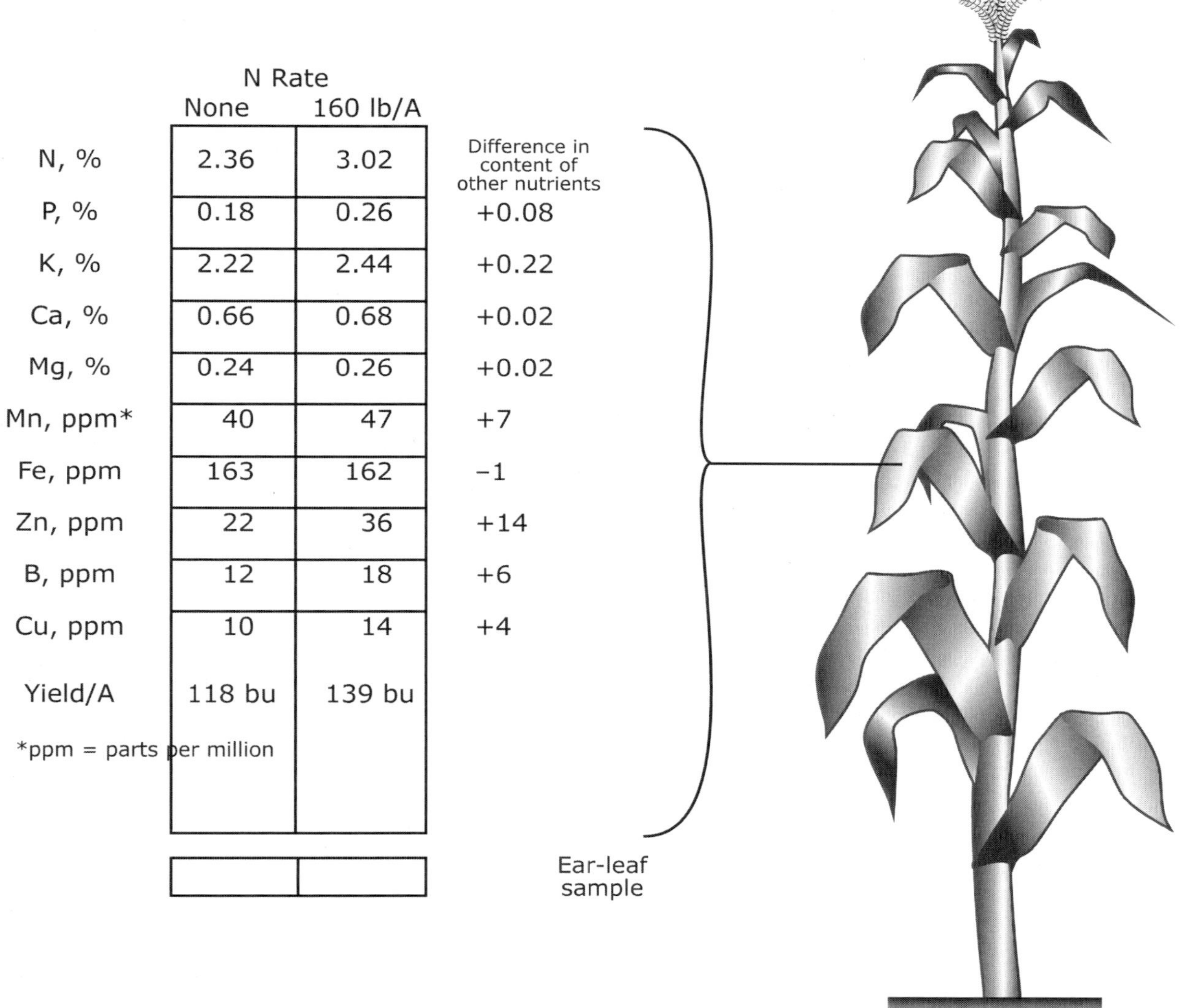

From University of Illinois data.

Tissue Testing

Sometimes the terms 'plant analysis' and 'tissue testing' are used synonymously, but the two are distinctively different. **Plant analysis** is a quantitative test done in a laboratory and is the determination of total nutrient content in plant tissue. **Tissue testing** is done in the field and is a qualitative, rapid colorimetric determination of the unassimilated, soluble nutrient content of tissue sap. It makes use of vials, test paper, color charts, and hand-held monitors and has the advantage of immediate results at a low cost. It is primarily limited to testing for nutrients in conductive tissue such as petioles, stems, or leaf mid-ribs.

A large amount of an unassimilated nutrient in the plant sap indicates that the plant is getting enough of the nutrient being tested for good growth. If the amount is low, there is a good chance that the nutrient is either deficient in the soil or is not being absorbed by the plant because of lack of soil moisture or some other factor.

Tissue tests can be run easily and rapidly in the field. Green plant tissue can be tested for several nutrients, including NO_3^-–N, P, K, and sometimes Mg, Mn, and Fe. However, it takes a lot of practice and experience to interpret the results, especially those for Mg and the micronutrients.

Tissue tests are used to identify one nutrient (N, P, or K) that may be limiting crop yields. If one nutrient is very low, others might "accumulate" in the sap because plant growth has been restricted, resulting in an improper interpretation. If the crop grows vigorously after the "deficiency" has been corrected, one might find that other nutrients are not present in amounts to produce high yields. What is identified or tested for is the most limiting nutrient at a particular growth stage.

On-the-spot tissue tests can be helpful in the hands of an expert. Without leaving the field, N deficiencies can be detected and corrective measure suggested. This savings of time could be valuable. As with total analysis of plants, it pays to compare healthy plants with poor ones whenever possible.

Kits containing instructions and supplies for running tissue tests are available. Many of them include test instructions and supplies for determining soil pH, NO_3^-–N, and soil P and K. Before using these tests, one should seek qualified training to develop diagnostic skills.

Field tissue tests for plants have not been used as widely as they might. It is simpler for most people to send in a soil or plant sample to a laboratory rather than to develop the skill to properly run and interpret these tests. Also, test supplies must be kept fresh and in proper working condition. When properly used, tissue tests work well with soil tests and plant analysis as another good diagnostic tool.

Soil Fertility Manual

Plant Deficiency Symptoms

For most of the last century, nutrient deficiency symptoms were common across much of North America. During the last quarter of the 1900s, however, improved fertilizer use removed soil fertility as a limiting yield factor in many fields. Toward the end of the century, intensive production and ever-increasing crop yields and less than optimum nutrient management helped to re-introduce nutrient deficiency symptoms in many areas.

The art of identifying nutrient hunger signs is important to profitable crop production. Many sources of information are available to those willing to develop their skills in identifying nutrient deficiency symptoms. They include textbooks, Extension bulletins, and charts that show full color examples. The Internet has added greatly to the availability of information on nutrient deficiencies. Research plots, Extension and industry demonstration plots, and farmer fields can help to 'calibrate' the eye of a trained diagnostician.

Table 9-2 can be a useful reference in identifying crop nutrient deficiencies. Some examples of nutrient deficiency symptoms and their location on the plant are shown in **Appendix A**.

> Review Questions:
> 24. Most crops start losing yield soon after visible nutrient deficiencies appear.
> 25. Nutrient deficiency symptoms of mobile nutrients appear first in (older; newer) leaf tissue?
> 25. Four nutrients translocated within plants are ______, ______, ______, and ______.

Importance of Cultural Practices

Knowing what has been done in a field ahead of time can be one of the most important diagnostic techniques to develop.

Get the facts:

- What crops have been grown?
- What tillage system was used?
- How was irrigation scheduled?
- How were the crops fertililized?
- When was aglime applied? What kind and how much?
- Have there been any other treatments? If so, what?
- When were the crops planted? Too early? Too late?
- Were insects, weeds, and diseases controlled?
- How did the crops look throughout the season?
- Was the weather too dry? Too hot? Too cold? Too wet?
- What were the yields?
- How was the quality of the crop?
- What are the yield goals?

Laboratories and agronomists will differ in their recommendations depending on answers to these questions, knowledge of grower managerial skill, yield goals, needs to increase soil test levels, and grower land tenure (the time the land will be operated by the grower). Building soil tests for higher yields is a capital investment that should be amortized over a period of years. (See discussion on long-term fertility build-up in Chapter 10).

The key to sound diagnosis and recommendations is to get the facts systematically and record them. A checklist will help in remembering important information. The diagnostic calendar shown in **Table 9-3** is a useful guide.

Table 9-2.   Key to nutrient deficiency symptoms in crops.

Nutrient	Color change in lower leaves (translocated nutrients)
Nitrogen	Plants light green; older leaves yellow (chlorosis), yellowing begins at leaf tip and extends along midribs in corn, sorghum
Phosphorus	Plants dark green with purple cast; leaves and plants small.
Potassium	Yellow/brown discoloration and scorching along outer margin of older leaves; begins at leaf tip in corn, sorghum.
Magnesium	Older leaves have yellow discoloration between veins, finally reddish-purple from edge inward.

Nutrient	Color change in upper leaves (nutrients not translocated)
	Terminal bud dies
Calcium	Emergence of primary leaves delayed; terminal buds deteriorate. Leaf tips may be stuck together in corn.
Boron	Leaves near growing point yellowed; growth buds appear white or light brown dead tissue.
	Terminal bud remains alive
Sulfur	Leaves, including veins, turn pale green to yellow, young leaves first.
Zinc	Pronounced interveinal chlorosis on citrus and bronzing of leaves. On corn, broad white to yellow bands appear on the leaves on each side of the midrib; plants stunted, shortened internodes. New growth may die in some bean species.
Iron	Leaves yellow to almost white; interveinal chlorosis to leaf tip.
Manganese	Leaves yellowish-gray or reddish gray with green veins.
Copper	Young leaves uniformly pale yellow; may wilt and wither without chlorosis. Heads do not form or may be grainless on small grains.
Chloride	Wilting of upper leaves, then chlorosis. In some cereal varieties, small chlorotic lesions on old leaves, progressing to younger leaves.
Molybdenum	Young leaves wilt and die along margins; chlorosis of older leaves due to inability to properly utilize N.

Remember:	Deficiency symptoms are not often clearly defined. Masking effects from other nutrients, secondary diseases, or insect infestations can prevent accurate field diagnosis.
Remember:	Deficiency symptoms always indicate severe starvation, never slight or moderate starvation.
Remember:	Many crops start losing yields well before deficiency signs start showing. This costly yield-limiting condition is called hidden hunger.

Hidden hunger may greatly reduce yields and quality without the crop ever showing any deficiency symptoms. More and more fields are suffering from this condition.

Table 9-3.   An example of a diagnostic calendar.

Start to Plan Ahead *But look back to* Population Yields Poor areas Drainage problems Crop lodging Compaction	**Fall**	• Sample soils under growing crops for next year. What is deficient? • Enough plants? Ears filled? Right size? Was yield goal high enough? • Note and map poor areas. Short growth? Poor stands? Barren stalks? Lodging? Deficiency symptoms? Weeds? Insects? Diseases? • Fall is best time to install drainage systems. Enough drains? • Root lodging: Root worms? Low K? Saturated soil? • Stalk breakage: Stalk borers? Diseases? Low K? Low Cl?
Set Yield Goals: Visit neighbors Visit dealers and advisers Attend meetings Read and study Check yield trials **Get Ready Early:** Machinery ready? Compaction potential? Less tillage? Fertility level? Drainage?	**Winter**	• What is a reasonable yield goal for the farm? For specific fields? • Compare notes with other leading growers • Secure supply of fertilizers, aglime, seed, and pesticides. • Study new information carefully: from research, Extension, agribusiness. • Read farm papers, technical journals, commercial releases. • Compare various hybrid sources, yield tests and long-time trends. Will hybrids respond to high populations and high fertility? • Planter or drill ready? Calibrated to plant optimum population? • Right seeding depth? Correct starter placement? • Avoid heavy equipment on wet soils. • Plan for conservation tillage, if possible. • Sample soils early. • Spot drainage problem areas during early rains. Ponding? Why? Cure?
Get Set! Go! Check: Emergence Seedling damage Cultivation damage Nitrogen needs: re-check! Weather patterns	**Spring**	• Plant early to take advantage of light, heat units. • Crusting? Planting depth? Germination? Temperature? Moisture? • Insects? Diseases? Chemical damage from herbicides or fertilizers? Birds? Shallow rooting? Saturated soil? • Root pruning? • Adequacy of N may depend on amount applied, rainfall, and yield potential. How much supplemental N is needed? • Chart stress periods throughout growth cycle. Record unusual weather during early growth, pollination period, and grain-filling stages.
Walk Fields! Insects? Diseases? Weeds? Nutrient hunger? Tissue quick tests Plant analyses Dig! Too dry? Compaction?	**Summer**	• Inspect fields several times during growth. Take soil map along. • Watch for diseases, insects, or obvious damage on leaves, stalks, ears, heads, pods, or roots. Are weeds robbing nutrients and water from crop? • Any known or unknown deficiency symptoms? • Use tissue tests, especially to detect hidden hunger or help explain growth differences between areas. • Dig deep to examine roots for restricted growth or discolorations. Why? Compaction? Drowned? Low K? Low Cl? Insects? Diseases? • Too dry? Irrigation scheduling adequate?
Look Again! Estimate yields **Check Harvest Equipment** **List Management Changes Needed**	**Fall**	• A ripe ear on a green corn stalk indicates adequate fertility and good hybrid. • Before and during harvest, ask: Was stand adequate?" • Always check yields accurately in some manner. • Do not leave yield in the field. Set combine properly. • Study checklist thoroughly, then summarize on back page. Test soils before seeing advisers.

Soil and plant analysis can help solve plant growth problems. Other diagnostic and information tools, such as the following, should not be excluded.

- **Printed material**—Books, pamphlets, visuals, and other information sources identify deficiency symptoms, diseases, insects, adapted varieties, etc. They are available at Extension offices and through fertilizer dealers, consultants, and other sources.

- **Extension specialists**, Certified Crop Advisers (CCAs), industry agronomists, company representatives, and consultants are in the field to help diagnose crop production problems.

- **Diagnostic workshops**—Many Extension services, as well as state and provincial associations, sponsor classroom training and field tours on soil and plant analysis, deficiency identification, facts about new varieties and hybrids, disease and weed control, soil conditions and tillage, and on other management practices.

- **Field days**—Research and demonstration field days sponsored by university and industry specialists provide great opportunities to study production practices in action.

- **Soil fertility short courses**—These help in the review of basic concepts and allow students to pick up new production techniques and ideas.

- **The Internet/websites**—The Internet has become an important source of information for all areas of crop production. Learning to use this powerful tool allows the diagnostician to explore a world of learning.

Putting It Together

Develop good diagnostic techniques. Learn and improve diagnostic ability and look beyond soil fertility problems (see **Table 9-4**). The diagnostic tools are available, but must be used. They include...a sampling probe...sample boxes for soil analysis...bags for plant analysis...directions for taking soil and plant samples...field information sheets...a shovel...a knife...plant tissue testing and pH kits...a notebook and pencil...a long tape measure...a hand lens...a checklist...a camera—preferably digital...cell phone...a global positioning system (GPS) receiver.

Most dealers recognize that diagnostic work is an important service to customers. It promotes good relations and profitable yields. Yet, too few growers systematically study their fields. Lack of motivation could be one reason. But more important is the idea of being qualified to diagnose the situation in a field. Practice, training, and calling on the experts will help remedy that.

Every field has something limiting its yields. Correct that limiting factor and yields will increase. Identify and correct the next limiting factor for another yield boost. The diagnostic process is a continuous challenge.

To be a compete diagnostician, one must look beyond fertility problems and understand environmental conditions and their effects on the crop. Such knowledge can help pinpoint a problem that is inducing or magnifying an apparent nutrient shortage. All factors that influence crop growth, response to fertilization, and yield should be evaluated.

- **Root zone**—The soil must be of good tilth and permeable enough for roots to expand and feed extensively. A crop will develop a root system 6 feet or more in depth in some soils to get water and nutrients. A shallow or compacted soil does not offer this root feeding zone. Wet or poorly drained soils result in shallow root systems.

- **Temperature**—Cool soil temperatures slow organic matter decomposition. This lessens the release of N, S, and other nutrients. Nutrients are less soluble in cool soils, and that increases deficiency potential. Phosphorus and K diffuse more slowly in cool soils. Root activity is decreased.

- **Soil pH**—Acid soil conditions reduce the availability of Ca, Mg, S, K, P, and Mo...and increase the availability of Fe, Mn, B, Cu, and Zn. Nitrogen is most available in the pH range of 6.0 to 7.0.

- **Insects and nematodes**—Don't mistake damage from these pests for deficiency symptoms. Examine roots, leaves, and stems for damage that may look like or may induce a nutrient deficiency.

- **Diseases**—Close study will show the difference between plant disease and nutrient deficiency. The disease can often be detected with a small hand-held lens, or magnifying glass.

- **Moisture conditions**—Dry soil conditions may create deficiencies for nutrients such as B, Cu, and K. This is why crops respond so well to these nutrients when they are available in dry periods. Drought slows movement of nutrients to the roots.

- **Soil salinity problems**—Soluble salts and alkali spots are problems in some areas, even though they may cover only part of the field. They are usually present when a high water table exists, where salt water contamination has occurred, or where poor water quality has been used for irrigation.

- **Weed identification**—Herbicides and mechanical controls are more important today than ever before. Weeds rob crop plants of water, air, light, and nutrients. Some weeds may even release substances that inhibit crop growth. Learn to identify weeds and to control them.

- **Tillage practices**—Some soils develop hardpans (compaction) and require deep tillage. This calls for more P and K to build up fertility. Also, it is desirable to know the fertility level of the subsoil.

- **Plant spacing**—Row width, spacing of plants in the row, and number of plants per acre have important effects on yields.

- **Water analysis**—Irrigation water can contain NO_3^--N, sulfate (SO_4^{2-})-S, B, K, bicarbonate, Cl, and other salts. A water analysis should be used to modify production practices for utilization of various water sources.

Summary

This chapter identifies soil testing as a major tool in high yield farming. But, a soil test is only as good as the sample. The benefit comes when soil tests are used with all other available information to help make recommendations for greater yields and profits. The person making the recommendations needs all available facts about the field and farmer.

Plant analysis is a tool to complement soil testing. But one does not replace the other. As with soil testing, good sampling is essential. Plant analysis data should be interpreted by scientists trained in this work.

Quick field tissue tests use the plant sap to check whether one element or multiple elements are limiting yields at that particular growth stage. They can be successfully performed in the field by a trained individual.

The deficiency symptoms of 13 nutrients were described. These symptoms are not often clearly defined. When they show, they mean severe starvation. Many crops start losing yields well before hunger signs develop. This is hidden hunger, a condition that drags down the crop's quality and yield before visual signs appear.

One must look beyond fertility problems to be a complete diagnostician. Consider the total environment, from root zone to tillage practices. The importance of cultural practices cannot be overstated when diagnosing crop conditions—every step from planting date through fertilizer and liming practices.

People talk about fine-tuning crop management. Being a good diagnostician is one way to keep valuable crops aimed toward higher yields. Soil testing, plant analysis, and diagnostic techniques should be used together to improve crop yields and profits.

Chapter 10

Fertilize for Profits

Many advances in crop fertilization practices have been achieved in past decades and will continue as farmers seek the most profitable and efficient methods. Fertilizer accounts for a significant share of total crop yield. Higher yield and potential profit per acre are closely related. The benefits of higher yield goals also include environmental protection and higher profits. Surprisingly, the market price for crops or the cost of fertilizer inputs may affect optimum application rates very little. Increasing soil test levels to optimum for phosphorus (P), potassium (K), and some other nutrients makes good sense. Building and maintaining soil fertility has been shown to increase long-term profitability. Positive interactions of nitrogen (N), P, K, and other nutrients should not be overlooked. Higher yield levels maximize the profit potential. Site-specific nutrient management is becoming a useful approach in many areas. Other management factors such as liming, reduced tillage, and quality benefits of fertilization should not be overlooked for their contributions to profitable returns.

Introduction

In 1950, total world fertilizer production was slightly more than 14 million tons; today, it is about 140 to 150 million tons. The remarkable...10-fold...increase in fertilizer production is well correlated to several factors, including increased world population and higher crop yields per unit of land. Also, there is greater awareness of the importance of sound fertilization practices...despite an increasing concern as to the perceived detrimental effects commercially produced fertilizer might have on the environment (see Chapter 11).

Beginning in the decade of the 1950s, other changes have taken place which helped to bring the production and use of fertilizer to the present. Some of the major ones include:

- Tremendous increases in the production and use of granular fertilizers, particularly in dry bulk blends;

- Development and growth of fluid fertilizers, largely made possible by the production of superphosphoric acid and ammonium polyphosphates;

- Soil application of anhydrous ammonia (NH_3) and other liquid fertilizers;

- Introduction of slow release materials such as urea formaldehyde and sulfur (S)-coated urea;

- Discovery and development of potash reserves in Canada;

- Development of phosphate deposits in North Carolina and the western U.S.;

- Placement methods and timing of fertilizer application;

- Growth of conservation tillage and residue management;

- Fertilizer management practices which optimize crop production while minimizing its effect on the environment.

- The adoption of site-specific crop production, including variable-rate fertilizer application.

In order to meet food demands of an ever-increasing world population, crop yields will need to continue to increase, perhaps doubling during the next 30 to 40 years. If this occurs, farming to feed North Americans and their world neighbors can be done on fewer acres. The more environmentally sensitive land can be placed under permanent vegetative cover, developed as wildlife preserves, or set aside for recreation.

The fact is clear that fertilizer use will continue to grow in importance as the world produces more people to feed. How to use this vital crop production input in a profitable, environmentally safe manner must continue to be researched as well.

Fertilizer and Farmer Profits

The only way farmers can stay in business—to sustain themselves and their families—is to make decent profits. Profitability is a logical reason, then, why farmers fertilize their crops. Consider what cutting back on fertilizer will and will not do.

- Cutting back on fertilizer will not cut land taxes, interest rates, seed and pesticide costs, machinery costs, or fuel costs.

- Cutting back on fertilizer below optimum levels will cut yields per acre, crop resistance to drought, disease, insects, and other stresses, and cut profits.

Fortunately, profitable crop production, including efficient use of fertilizer and other best management practices (BMPs), and environmental protection go hand-in-hand. **Table 10-1** illustrates how N fertilization and plant population management, both BMPs, interact to increase corn yields and, thus, potential farmer profit. At the same time, less N remains in the soil, so the potential for nitrate-N (NO_3^--N) finding its way to groundwater is reduced (data not shown).

Table 10-1. Increasing corn plant population interacts with higher N to boost yield and efficiency.

Population, plants/A	Yield, bu/A, at N rate of:			Response to N, bu/A
	80 lb/A	160 lb/A	240 lb/A	
12,000	118	138	155	37
24,000	151	178	202	51
36,000	164	210	231	67
Response to population, bu/A	46	72	76	

Florida

Fertilizer accounts for as much as one-third or more of total crop yield. On many high-yielding fields, increased production from fertilizer can be 60% or higher. In many cases, optimum fertilization represents the difference between a substantial profit and a loss. Many believe no other production input brings a greater return than efficient fertilizer use...returning $3, $4, $5, or more for each $1 spent.

Aim for a Higher Yield

There are several components of farmer profits.

Production costs—The farmer can do little to control rising costs of production, except to employ BMPs to ensure greater efficiency of inputs.

Selling price—The farmer can optimize price received for a crop through intelligent marketing, but has no control over markets, except under certain local conditions of supply and demand.

Crop yield and quality—The farmer can increase yield and improve quality. High yields of high quality crops should be a primary objective.

Figure 10-1 illustrates the concept of maximum economic yield (MEY).

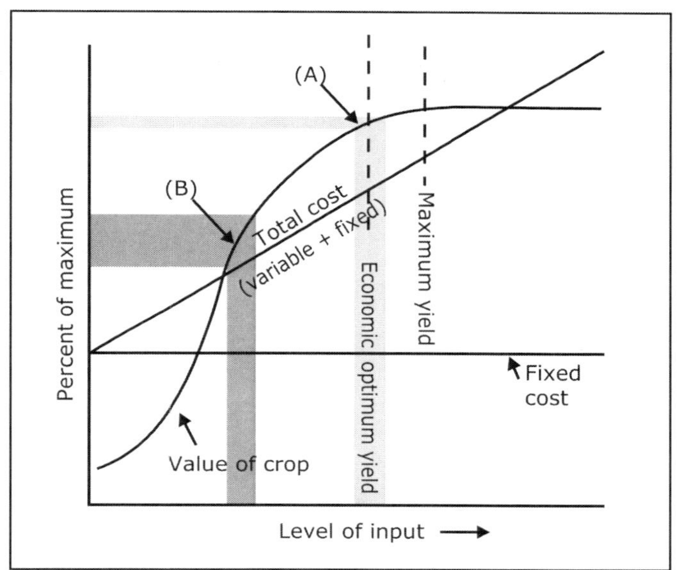

Figure 10-1. Typical response of crop to an input such as fertilizer and generalized relationship to costs and production.

The light gray band (A) represents the effect of reduction of input (such as fertilizer) when managing at a point near economic optimum...MEY. This is usually within the range where maintenance-only application rates are needed. The dark gray band (B) represents the effect of the same reduction of input when managing at a point where buildup applications are needed. Reduction of input in this zone in this example results in yield losses of several times the loss caused by the same input reduction when managing in the zone near economic optimum. Managing for high yields, near the MEY level, provides the greatest flexibility in management decisions.

Review Question:
7. Four components of farmer profits are _______ _______, _______ _______, _______ _______, and _______ _______.

Setting Yield Goals

Yield goals should be set for each production field. They should be both realistic and challenging. Start with an analysis of past yields. Look at weather patterns and evaluate management decisions and inputs that can be improved or changed.

Establish a 3- to 5-year yield improvement program, with a goal of increasing yields by 10 to 20% per year. The percentage increase will depend on several factors, including previous yields and crop(s) being grown. Also, when dealing with larger acreages, a more conservative approach should be taken; a more aggressive one with smaller acreages. Look for the weakest practice, improve it, and upgrade other management inputs at the same time.

Table 10-2 shows how P and K increased wheat yields, reduced disease, and improved N use efficiency.

Table 10-2. Interaction of P and K can increase wheat yield, reduce disease, and improve N use efficiency.

P_2O_5 rate, lb/A	0 K_2O		40 lb/A K_2O	
	Yield, bu/A	Disease, %	Yield, bu/A	Disease, %
0	40	48	51	40
30	61	65	76	40
60	61	57	84	35

N rate @ 75 lb/A　　　　　　　　　　　　　　　　　　　Kansas

Data in **Table 10-2** illustrate the importance of balanced fertilization in boosting yields while increasing efficiency of purchased inputs.

Moving up the yield goal ladder should be a continuous management process. That is, when the goal is achieved, a new and higher mark should be set.

High Yields: Environmental Protection, Lower Unit Costs, and Higher Profits

High yields protect the environment, an important factor in the sustainability of production agriculture. Vigorous, early plant growth produces quicker ground cover (canopy) to protect soil from water and wind erosion. Plants develop more robust root systems to hold the soil in place and allow it to absorb rainfall more quickly. Plants use nutrients and soil water more efficiently and produce larger quantities of crop residues, to further protect against the ravages of water and wind erosion. When decomposed, these residues also recycle nutrients and organic matter back into the soil. **Table 10-3** shows the relationship among residue levels, runoff, and soil loss.

Table 10-3. Effect of surface residue on runoff and soil loss.

Test	Residue, tons/A	Runoff, as a % of rainfall	Soil loss, tons/A
1	0	45	12
2	0.25	40	3
3	0.5	25	1
4	1.0	5	0.3

Indiana

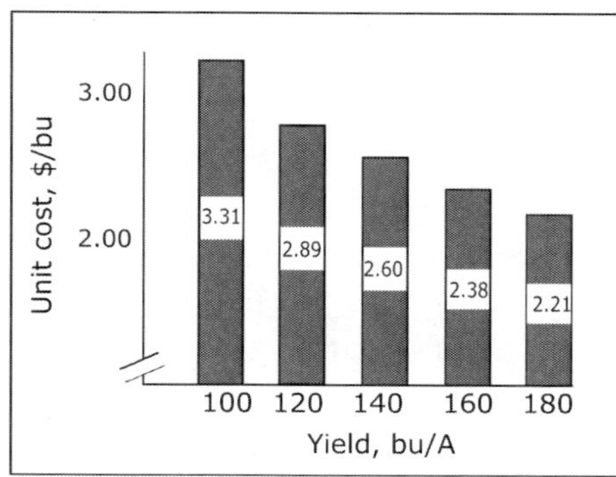

Figure 10-2. Higher corn yields help to lower unit production costs. (Illinois)

Figure 10-2 shows the relationship between corn yield and unit production cost. As yield increased from 100 to 180 bu/A, cost per bushel dropped from $3.31 to $2.21, a difference of $1.10.

Table 10-4 shows how a combination of yields and low unit costs can be projected across a 300-acre corn farm, using the data in **Figure 10-2** and two corn prices. Numbers in parentheses represent losses.

Table 10-4. Influence of yield and unit production cost on profitability; 300 acres, two corn prices.

Corn price, $/bu	Profit on 300 acres for [yield (bu/A)/cost ($)]				
	100/3.31	120/2.89	140/2.60	160/2.38	180/2.21
2.50	(24,300)[1]	(14,040)	(4,200)	5,760	15,660
3.00	(9,300)	3,960	16,800	29,760	42,660

[1] (Losses in parentheses)

The beneficial interaction between yield and unit cost as shown in **Table 10-4** is obvious. Well managed, high yielding crop production reduces risks associated with low prices and broadens the window of opportunity for profits. It follows that a sound marketing plan opens the window even further.

The principles illustrated by **Figure 10-2** and **Table 10-4** apply to other crops as well as corn. **Table 10-5** shows how higher yields and lower unit costs translate into higher profits on alfalfa.

Table 10-5. Lower unit costs mean higher profits from alfalfa.

Average yield, tons/A	Production costs		Net profit, $/A[1]		
	$/A	$/ton	$80/ton	$95/ton	$110/ton
3.5	260	74	20	72	125
4.6	242	53	126	195	262
6.6	292	44	236	335	434
8.2	313	38	343	466	589

[1]Assuming average U.S. cost of $100/t   Pennsylvania

The above examples may not fit on individual farms or in different areas...but the principle does. So do the results. As yields go up, unit costs drop, and profits per acre increase.

Review Questions:

9. (T or F) High yields help to protect the environment

10. (T or F) High yields lower unit production costs and give a higher per acre return.

11. (T or F) As unit production costs drop, profits go up.

Price of Crop or Price of Fertilizer Affects Optimum Rate Very Little

The optimum rate of fertilizer is changed very little by price, either crop or fertilizer, as long as the crop continues to be responsive to the nutrient being applied. Examples illustrating this principle are shown in **Table 10-6** and **Table 10-7**.

Table 10-6. Optimum N rates change little with corn and fertilizer price fluctuations.

Corn price, $/bu	Optimum N rates on corn, lb/A		
	N cost, $/lb		
	0.12	0.18	0.24
2.00	182	174	166
2.50	189	180	172
3.00	192	184	176

When N costs are highest and corn prices are lowest, optimum N rate is 166 lb/A. At the highest corn price and lowest N price, optimum rate is 192 lb/A, an increase of only 26 lb or 15.7%. At the same time, the N cost:corn price ratio changed by a factor of 3.

Table 10-7. Optimum K_2O rates change little with corn and fertilizer price fluctuations.

| Corn price, $/bu | Optimum K_2O rates for corn, lb/A | | |
| | K_2O cost, $/lb | | |
	0.11	0.13	0.15
2.00	115	105	95
2.50	125	115	105
3.00	135	125	115

Although corn prices dropped from $3.00 to $2.00 per bushel, optimum K_2O rates dropped only 20 lb/A at each of the three K_2O prices. Even at the higher K_2O price and $2.00 corn, optimum K_2O rate was still 95 lb/A.

When the data in **Table 10-6** and **Table 10-7** are analyzed, two facts become evident.

- Crop prices affect optimum fertilizer rates much less than most people believe. Why? Because fertilizer represents a relatively small percentage of total production costs, while the returns are high when plant nutrients are applied and used efficiently.

- Even if fertilizer prices increase significantly...always a concern among farmers...there is little or no economic justification for drastic cutbacks in nutrient use.

The message is loud and clear in both cases. Fertilizer rates should be maintained as indicated by soil tests, even with lower crop prices and/or higher fertilizer costs. A majority of agronomists believe the most profitable fertilizer rate turns out to be the one near the top of the yield response curve.

Soil tests should be built into the optimum range and maintained to support top yields, lower unit production costs, and increase profit potential.

Building Soil Tests: A Long-Term Investment

Nitrogen fertilizer purchases represent a short-term investment because the majority of the return is expected during the year of application. Nutrients such as P and K are a different story, however, because only a fraction of the total return from their use is ordinarily realized during the first year. In a majority of soils, much of the P and K remains in forms available for future crop uptake.

Just as the costs of installing field drains or an irrigation system are recovered over several years, so are the costs of P and K...which should be amortized over several years. Benefits of P and K are long-term in nature and should be treated as such.

The example illustrated in **Table 10-8** shows the relatively minor cost of building a soil P test from 45 to 55 lb/A and the amount of yield response necessary to pay that cost. The buildup would require about 90 lb P_2O_5/A at a cost of $24.30, assuming a P_2O_5 price of 27¢/lb.

Table 10-8. Annual payment or corn yield increase necessary to pay off the cost of a 10 lb/A P soil test buildup at 8% interest.

Payoff period, years	Required annual payment, $/A	Required annual yield response, bu/A, if corn price/bu is:		
		$2.00	$2.50	$3.00
1	26.24	13.1	10.5	8.7
5	6.00	3.0	2.4	2.0
10	3.62	1.8	1.4	1.2
20	2.48	1.2	1.0	0.8

The above example shows that, for a 10-year payoff and $2.50 corn, yield would need to be increased by only 1.4 bu/A per year to recover the cost of fertilizer P_2O_5. Or, if no response occurred in five of the 10 years, an average 2.8 bu increase during the other five would still pay for the P_2O_5. It doesn't make economic sense to allow P, K, and other essential nutrients to limit yields.

Fertility buildup offers several advantages to the farmer, whether as the landowner or tenant.

- Once soil test levels are built to the optimum range, then maintenance applications prevent soil depletion. Yield potential is maintained for five, 10, even 50 years or more.

- Building soil fertility will return higher yields each year of amortization, even with drought, excess moisture, cold, heat, disease, insects, and other stresses.

- Land value (resale) goes up with productivity. Smart buyers will pay more for well managed land because they know their out-of-pocket costs for fertilizer will be less while yield potential remains high.

Remember that high yields remove large quantities of the primary and secondary nutrients and can deplete the soil supply of micronutrients in some cases. (See **Appendix B-1**.) In addition to removal by the crop, nutrients can be lost by soil tie-up, erosion, and volatilization from soil and leaf surfaces.

A key point is to check soil test levels regularly, to be sure they are being built to and maintained at high levels.

Long-Term Effects of Fertilizer Use

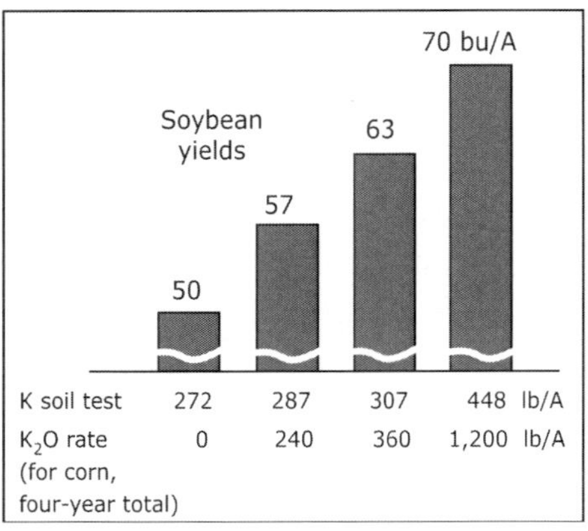

Figure 10-3. Potassium applied for corn builds soil tests and boosts soybean yields. (Ohio)

Building and maintaining soil fertility is an important part of long-term profitability. Yields and profits increase as farmers improve management, including fertilizer practices, over the years. **Figure 10-3** shows how building soil K levels during a four-year continuous corn cropping sequence increased soil test values and boosted soybeans grown after the fourth corn crop. The residual benefits of high soil K fertility on soybean yields are only part of the story, however. The average annual increase in corn yields during the four years was 25 bu/A/yr.

The farmer must keep in mind that basing fertilizer management decisions on short-term results can be misleading, even disastrous. **Table 10-9** illustrates the point well.

Table 10-9.   Long-term response to P_2O_5 fertilization.

Five-year sequence	Response to 40 lb P_2O_5/A/yr[1]	
	bu/A/yr	Profit, $/A/yr
1st five years	1	(7)[3]
2nd five years	21	42
3rd five years	41	92
4th five years	47	107
5th five years	34	75
6th five years	74	175
7th five years	80	190
8th five years[2]	69	162

[1]N applied at 160 lb/A/yr   Kansas
[2]One year lost because of hail
[3]() indicates loss
Low soil P

Profits shown in **Table 10-9** are based on $2.50 corn and P_2O_5 at 25¢/lb. During the first five years, the return from P fertilization would have been a negative $35/A ($7.00 x 5). If the management decision had been made to eliminate P after that period, total revenue loss for the 40-year sequence would have been $4,180/A, based on averages of the eight five-year periods. That is an average loss of nearly $120/A/yr above P costs for each of the 35 years following the decision to stop using P. For a quarter section of land (160 acres), that amounts to $19,100/yr less returns, or a total loss of $669,000 over the last 35 years.

Interactions and Fertilizer Efficiency

An interaction refers to the effect of one production input or yield factor on the response to another factor (see **Table 10-1**). **Table 10-10** illustrates the effects of planting date on K fertilization. On this low K soil, K and planting date interacted to boost corn yield. At the early June planting date, K efficiency was less than half as much as when corn was planted by mid-May. The significance of this example is that fertilizer efficiency is optimized only when other sound management principles are followed.

Data in **Table 10-11** show how N, P, and K interacted to boost wheat yields and increase N use efficiency. Analysis of the data shows that P was the most limiting of the three nutrients, but that it took both P and K to achieve the highest yield.

Table 10-10. Potassium and planting date interact to boost corn yield.

Planting date	Increase from K_2O, bu/A
Early April	26
Early May	25
Mid May	20
Early June	12

Low soil K

Table 10-11. Interaction of N, P, and K can increase wheat yields and improve N use efficiency.

Treatment, lb/A			Grain yield, bu/A	N efficiency, bu/lb
N	P_2O_5	K_2O		
0	0	0	29	–
100	0	0	24	0.24
100	120	0	45	0.45
100	0	120	26	0.26
100	120	120	52	0.52

Oklahoma

Crops show greater response when fertilization and other management interact positively. Interactions include factors such as row spacing, planting date, weed and other pest control, variety, rotation, soil pH...and others. Best returns from fertilizer are realized when a BMP production system is followed. Many interactions that influence fertilizer use efficiency involve management practices which cost little or nothing (**Table 10-12**). Timeliness, for example, is most important whether it is planting on time, having everything 'ready to go' at the appointed time, controlling pests, or observing fields.

Table 10-12. Some management practices that cost little or nothing extra.

• Timeliness	• Tillage
• Variety, hybrid	• Planting date
• Row width	• Seed drop
• Fertilizer placement	• Record keeping
• Scouting	• Population

Fertilizing for Maximum Economic Yield

Maximum economic yield (MEY) is that yield where unit costs are lowered to the point of highest net return per acre...the most profitable yield.

The MEY varies from year to year and from field to field. Weather, pests, and other yield-influencing factors must be dealt with on a site-specific basis. Each will affect the management of production inputs, including fertilization.

When does it become unprofitable to increase fertilizer rates? Certainly not at the point of maximum return per dollar invested...rather, at the point of maximum profit. Much is said about diminishing returns. The key to additional fertilizer increments is not whether the last increment produced a return as great as the one preceding it, but whether the return was greater than the cost. **Table 10-13** illustrates the principle.

Table 10-13. Returns from N fertilization of corn.

N rate, lb/A	Yield, bu/A	Increase, bu/A	Net return for N, $/A	Return/$ of N
0	79	—	—	—
30	100	21	46.50	7.75
60	117	17	36.50	6.08
90	131	14	29.00	4.83
120	142	11	21.50	3.58
150	150	8	14.00	2.33
180	154	4	4.00	0.67
210	155	1	(3.50)	(0.58)

Based on $2.50 corn and an N price of 20¢/lb   Illinois

The 30 lb increment from 120 to 150 lb/A of N produced less than a third of the return from the initial 30 lb of N, but still returned a net of $14.00/A, or $2.33, for each dollar invested. The farmer should weigh the risks—how much margin against potential lost yield or low crop prices—he or she is willing to take. In the end, though, the best chance for highest sustained profits comes with management geared to MEY.

Earlier in this chapter, profit windows were related to high yields and low unit production costs. Top farmers set high yield goals because they want to maximize their profit window or 'profit zone'. They know that things don't always go as planned.

- Yields can be lower than goals set;

- Prices can be lower than expected;

- Both yields and prices can be lower than expected;

- Production costs can exceed amounts budgeted.

High, efficiently produced yields can help offset these negatives and expand profit zones. **Figure 10-4** shows three profit zones, three target yields, and three levels of production costs for $2.00, $2.50 and $3.00 corn. Note that profit zones expand as yields increase...as MEY is approached.

A farmer with a yield of 70 bu/A of corn could tolerate only a 7 bu loss or a 36¢/bu drop in price without suffering a loss, even with $3.00 corn. On the other hand, the farmer with a 180 bu goal could still make money with reductions of 69 bu or $1.15 in price. The answer is clear: Fertilize and follow other BMPs to achieve MEY.

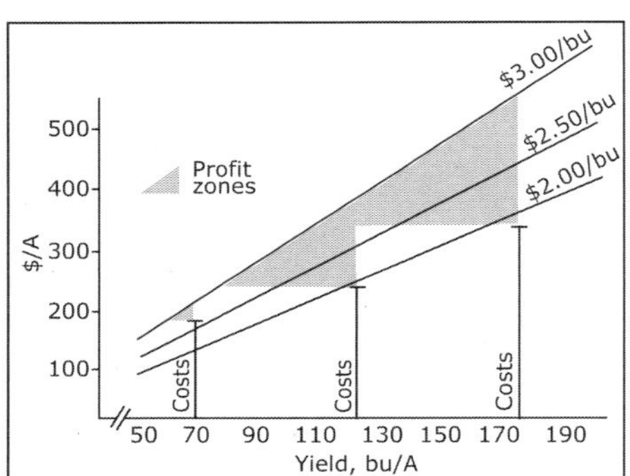

Figure 10-4. Profit zones as influenced by corn price, costs and yield.

Review Questions:
23. That crop yield where unit costs are lowered to the point of highest net return per acre is known as ______ ______ ______.
24. (T or F) MEY is a system of BMPs.
25. (T or F) The key to deciding whether to use additional fertilizer increments is whether or not the last increment produced a return as high as the previous one.
26. Profit potential zones ______ as yields increase.

Site-Specific Nutrient Management

While site-specific nutrient management has been extensively evaluated in recent years, some involved in crop production still have not fully grasped its potential applicability and value. The lack of understanding and acceptance of the concept is due to several factors, including:

- Difficulty in establishing soil test variability;

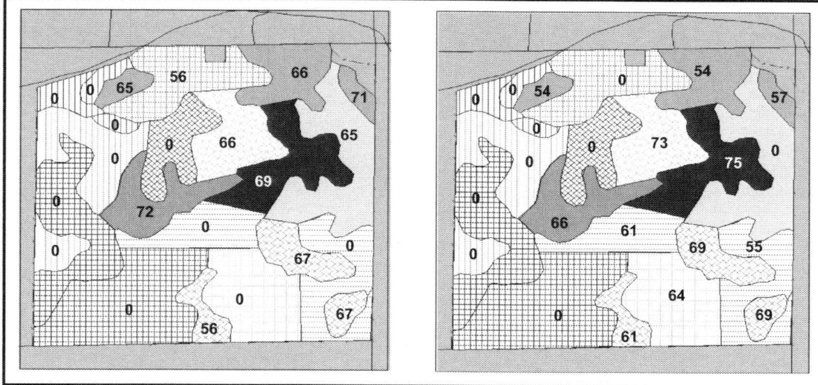

- Fitting current nutrient recommendations to changing production systems that involve higher yields and increasing use of conservation tillage systems;

- Accuracy of variable rate nutrient applications;

- Higher levels of management required to make the system work.

The following example illustrates how one approach, using zone management, varies from field-average management and has the potential to significantly alter nutrient use.

In a 157-acre field in central Indiana, 3-year average corn yields varied by soil type. Larger soil mapping units were subdivided into areas small enough for representative soil samples to be collected. They ranged in size from 1 to 19 acres. All samples were composites of 20 to 30 individual soil cores, each geo-referenced and taken to a depth of 6 in. These subdivisions in soil mapping units formed the basis for management zones.

Recommendations for P and K were calculated using yield goal, soil test level, cation exchange capacity (CEC), and soil buffer capacity. The 3-year average corn yields were used as the yield goal for each zone, which ranged from 138 to 213 bu/A. The Bray P-1 and Mehlich 3 K soil tests for each management zone ranged from 22 to 97 parts per million (ppm) and 111 to 279 ppm, respectively. Field-average P and K soil test levels, calculated from zone soil tests using zone area as a weighting factor, were 52 and 170 ppm, respectively.

Recommendations based on calculated field average would have called for no supplemental P or K. However, with more intensive sampling and yield goal assessment, approximately 73 acres (46%) were identified as needing additional P, and 76 acres (48%) required additional K. The recommended rates per acre for each management zone are shown in **Figure 10-5**. Total field requirements of 4,740 lb P₂O₅ and 4,895 lb of K₂O were identified by the management zone plan, but were missed by whole field sampling and yield goal determination.

This example illustrates the potential for site-specific management in refining nutrient use. The frequency at which disparities between site-specific and whole field nutrient management occur is not known. The degree of variation between the two concepts has not been established, but must be addressed

Review Questions:

27. (T or F) The concept of site-specific management is easily understood.

28. One of the difficulties of site-specific management is establishing
______ ______ ______.

on a field by field basis. However, what is known is that proper application of site-specific nutrient management can result in more efficient use of fertilizer and, thus, greater return on dollars invested.

Other Aspects of Fertilization:

• Lime and Soil pH

Liming acid soils to improve fertilizer efficiency is an important management input. It reduces levels of toxic substances in the soil, improves physical characteristics, and enhances microbial activity. **Table 10-14** shows how dramatic crop response to liming can be. How efficient would fertilization have been on such soils? (See Chapter 2 for more on liming.)

Table 10-14. Soybean response to liming.

Soil pH	Lime rate, tons/A	Yield response, bu/A
5.1	0	–
	1.0	20
4.2	0	–
	6.0	28

• Tillage

The shift to reduced tillage systems is forcing changes in methods of fertilizer application. Residue management, and other factors are altering rooting patterns and growth habits because of changes in soil temperatures, moisture retention, nutrient distribution, and organic matter buildup. Chapters 3, 4, 5 and 10 discuss various aspects of tillage and residue management on fertilizer use and environmental protection. Use of starter fertilization and placement become more important as tillage frequency is reduced.

• Improved Crop Quality

In many instances, the extra quality produced by fertilizers increases market value enough to pay for the fertilizer. Better plant nutrition influences quality in many ways. Balanced fertilization is important for proper maturity development, shown in **Production Concept 10-1**.

Nitrogen increases protein content of non-legumes, both grain and forage. **Table 10-15** shows how N increased both yield and protein content of corn.

Phosphorus also improves quality of grain and forage. In an Arizona study, beef cows with 0.22% P in their feed had a conception rate of 59%, compared to 89% when dietary P was increased to 0.3%.

Table 10-15. **Nitrogen increases yield and protein content of corn.**

N rate, lb/A	Grain yield, bu/A	Protein in grain, %
0	117	8.0
80	160	8.5
160	183	9.5

Potassium reduces intensity of crop diseases such as pod and stem blight of soybeans, leading to higher quality and less dockage (see **Production Concept 5-1**). Potassium greatly reduces leaf disease on Coastal bermudagrass and increases lint quality of cotton. Chloride (Cl) significantly reduces the intensity of many fungal diseases, particularly of small grains, advancing maturity and increasing yields.

Other essential plant nutrients can affect crop quality as well. **Figure 10-6** shows that S increased protein content in bahiagrass.

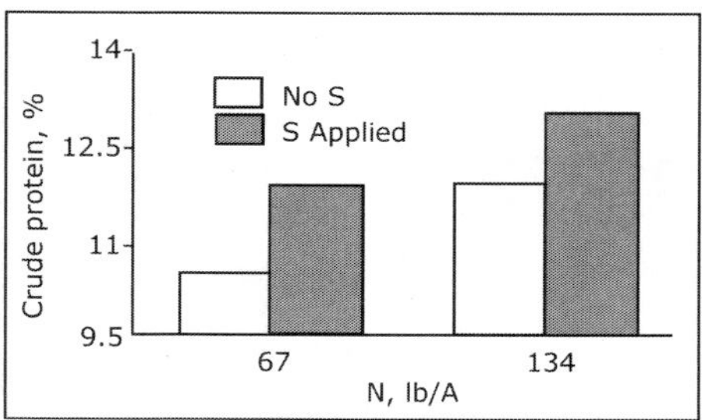

Figure 10-6. Sulfur increases protein content of bahiagrass. (Florida)

• Organic Fertilizer Sources

The use of organic, or natural fertilizer sources is receiving much attention in North America and around the world. While production agriculture has utilized organic sources such as animal manures, and continues to do so, there are practical and economic limitations to its use. Further, commercially produced fertilizers are as natural in their origin as are organic sources.

From a crop's view, there is no difference among the various nutrient sources because plants don't use nutrients in the original form. Rather, all nutrients must be in the ionic form before they can be taken up by the plant roots. So, the original source, whether it is organic or inorganic, has no identity to a healthy, vigorous plant. Plants do require the presence of nutrients in an adequate, continuous, and balanced supply to assure proper growth.

Concerns by some that commercial fertilizer is not natural (and thus, bad) have no scientific basis. Processing and utilization of P and K deposits represent the ultimate in recycling... moving nutrients back to the soil from which they were originally released by weathering. As for N, approximately 80% of the Earth's atmosphere is gaseous N, the same N we breathe with every breath and which is used to manufacture N fertilizers and which is fixed by legumes.

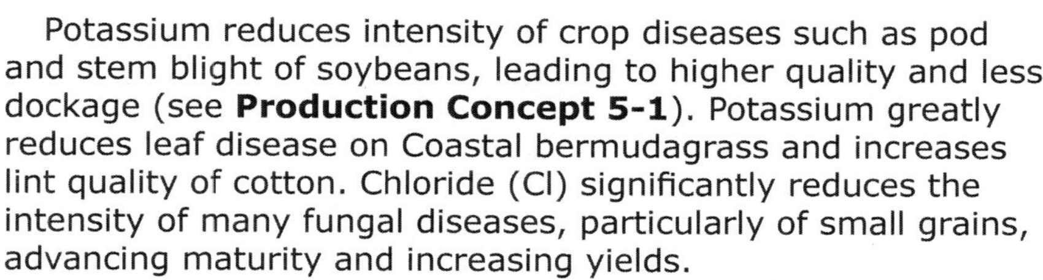

Review Questions:

29. (T or F) Liming reduces toxic substances in the soil.

30. (T or F) The shift to reduced tillage systems will have little influence on methods of fertilizer application.

31. Nitrogen (increases, decreases) protein content of non-legumes

32. Phosphorus increases P content of grain and forage and improves ______ ______ in beef cows.

33. Potassium reduces soybean diseases such as ______ and ______ blight.

34. (T or F) Sulfur increases protein content of crops.

35. There are ______ and ______ limitations to the use of organic fertilizer sources such as animal manure.

36. (T or F) From a crop's point of view, there is no difference among various fertilizer sources.

Fertilizer Speeds Maturity

Plant maturity is like farm profit when it comes to crop growth. Maturity, like profit, is in balance...FERTILITY BALANCE. The plant is there to grow, to reproduce itself...to produce seed. So, when the plant runs short on an essential nutrient, it slows down, maturity is delayed, and yield is often reduced as shown in the corn and cotton examples below. Proper...BALANCED...fertilization helps speed maturity and assure best yields and profits.

Effect on corn maturity... and yield

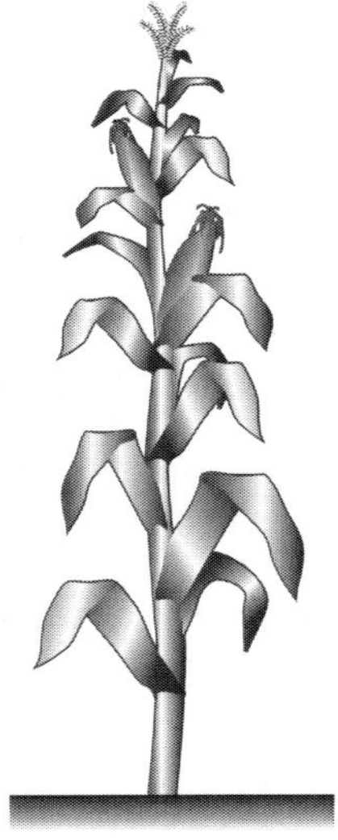

	Fertilizer rate, lb/A	
	$N-P_2O-K_2O$	$N-P_2O_5-K_2O$
	300-150-0	300-150-200
Silked on July 21, %	4	42
Yield, bu/A	151	188

Effect on cotton maturity... and yield

	Fertilizer rate, lb/A	
	$N-P_2O-K_2O$	$N-P_2O_5-K_2O$
	0-50-50	100-50-50
First pick, %	66	81
Lint yield, lb/A	1,896	2,089

Fertilizer's Place on the Whole Farm

Sometimes cash flow problems call for short-term returns rather than long-term benefits. In such instances, all potential input purchases, including fertilizer, must be evaluated in relation to the goal of maximizing return on the last dollar spent. **Table 10-16** illustrates this principle on the short-term return from the last increment of purchased P fertilizer. Soybean price was set at $5.00/bu, P_2O_5 at 25¢/lb.

Table 10-16. As P rate increases for soybeans, profit increases to a maximum, but short-term return per dollar invested on the last increment declines.

P_2O_5 rate, lb/A	Yield, bu/A	Added yield, bu/A	Added value, $/A	Fertilizer P_2O_5 profit, $/A	Return on the last increment, $/$
0	34.9	—	—	—	—
15	38.4	3.3	16.50	12.75	4.40
30	41.0	2.6	13.00	22.00	3.47
45	42.6	1.6	8.00	26.25	2.13
60	43.3	0.7	3.50	26.00	(0.07)

In the example in **Table 10-16**, a farmer wishing to maximize profit would apply P_2O_5 up to a rate between 45 and 60 lb/A. However, if capital is limiting all desired purchases, return to the last P increment becomes important. If another input, such as a protein supplement, is likely to return $2.00 per dollar invested, the supplement should be purchased before P_2O_5 fertilization exceeds 45 lb/A.

Summary

Remember, fertilizer management should be viewed in the long-term, but short-term economic reality sometimes takes precedent. It is emphasized, however, that such considerations are important only when short-term economics dictate. SFM

Chapter 11

Plant Nutrients and the Environment

Plant nutrients are vital to producing adequate food and fiber, but some can pose an environmental concern unless managed properly. Nitrogen (N) can be lost from the soil, resulting in both an economic and environmental risk. Phosphorus (P) also requires careful management to avoid undesirable effects to water bodies. Potassium (K), magnesium (Mg), sulfur (S), and micronutrients pose little or no threat to the environment and are important for efficient utilization of other nutrients. Management objectives should include balancing all production inputs at optimum levels and using site-specific soil and water conservation techniques. Properly developed nutrient management plans can help achieve crop production and environmental goals.

Introduction

All the essential nutrients required for food, feed, and fiber production are involved with the quality of our environment. Collectively, they enhance both the productive potential and environmental integrity of farm enterprises when used in adequate and balanced amounts.

Plant nutrients promote a more vigorous, healthy, and productive crop...one which develops greater root systems, more above-ground residue, quicker ground cover, greater water use efficiency, and higher resistance to crop stresses such as drought, pests, and cold temperatures.

Although the essential plant nutrients play a vital role in providing adequate food supplies and protecting our environment, some pose an environmental risk with improper management. The two nutrients most often associated with mismanagement and nonpoint source environmental concerns are N and P.

> Review Questions:
>
> 1. (T or F) Use of proper amounts of the essential plant nutrients enhances our environment.
>
> 2. The two nutrients most often associated with nonpoint source environmental concerns are ________ and ______.

Nitrogen

Nitrogen losses can occur with soil erosion or runoff. Soil N in crop residues, animal manures, and in other organic fractions in the soil (including soil microbes) is subject to surface erosion and movement with water and soil sediment.

Runoff of fertilizer N has been blamed for low dissolved oxygen (hypoxia) in the shallow waters of the Gulf of Mexico along the Louisiana-Texas coast. Hypoxia can cause death or stress in bottom-dwelling fish, shrimp, and other organisms that cannot move out of the hypoxic zone.

The Mississippi River Basin drains about 40% of the contiguous U.S. It includes 80% of U.S. corn and soybean acreage and much of the acreage of wheat, cotton, and other crops. Excessive N fertilization in the upper Mississippi River Basin states is frequently singled out and blamed as the cause of the hypoxic zone in the Gulf of Mexico. However, there are numerous N

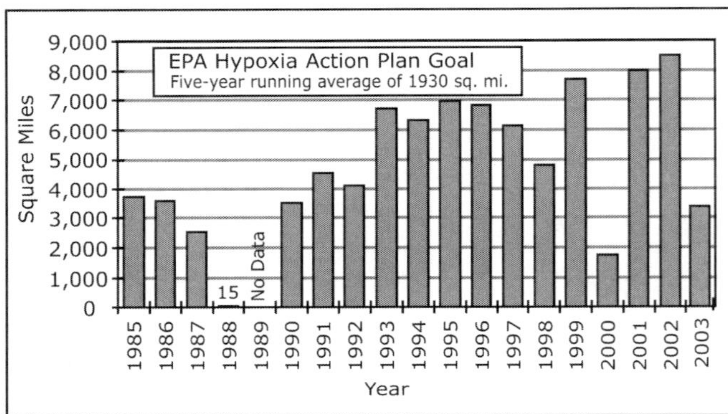

Figure 11-1. **Approximate size of the hypoxia zone in the northern Gulf of Mexico.**
Source: N. Rabalais, Louisiana Universities Marine Consortium.

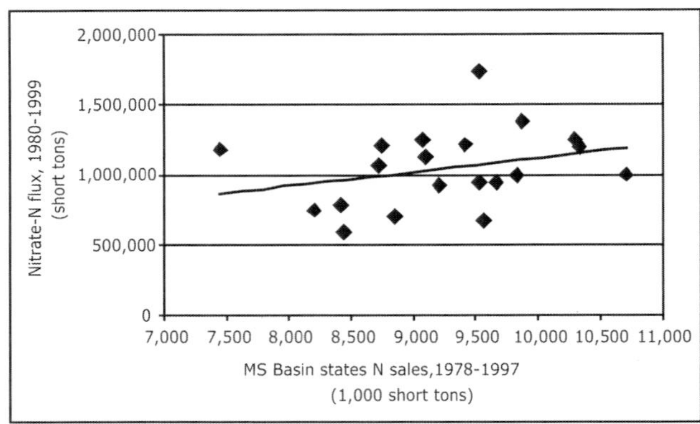

Figure 11-2. **Nitrate-nitrogen discharge to the Gulf of Mexico vs. two year lag in Mississippi River Basin fertilizer nitrogen sales.**
Source: U.S. Geological Survey (Don Goolsby) and The Fertilizer Institute.

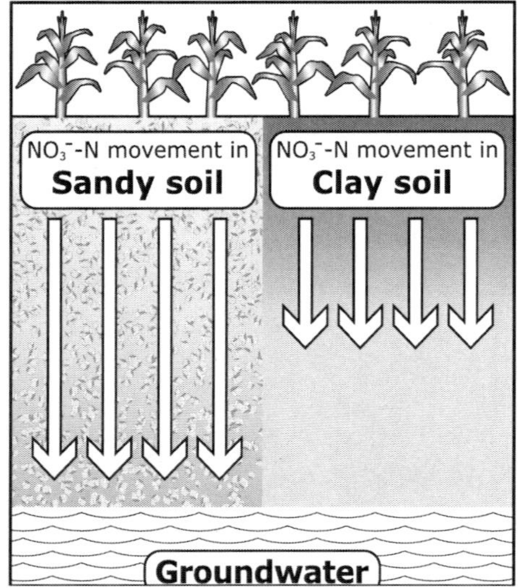

Figure 11-3. **Nitrate is more likely to move downward in sandy soil than in clay soil.**

Review Questions:

3. The form of N most often associated with groundwater and leaching losses is ________.

4. (T or F) The nitrate ion has a negative charge.

5. All N sources undergo soil transformations. These transformation are part of the total process known as the ________ ________.

6. (T or F) Nitrogen from animal manure and legumes is not subject to leaching losses.

sources in addition to fertilizer. Nitrogen input into the Mississippi River comes from soil organic matter, animal wastes, sewage sludge, legumes, atmospheric deposition, and runoff from urban sources (e.g. lawns, golf courses, athletic fields). Since 1980, nitrate-N ($NO_3^-–N$) delivery into the Gulf of Mexico has predominately been controlled by the volume of water flow and not N fertilizer use (**Figure 11-1** and **11-2).**

Hypoxia in the Gulf of Mexico is the result of complex interactions of chemical, biological and physical processes, and while N fertilization is one of many sources that may contribute to the problem, it is manageable concern that can be addressed through good farming practices and fertilizer management.

Much of the concern about N in the environment is due to the potential movement of unused or excess NO_3^--N through the soil profile into groundwater (leaching). Because of its negative charge, NO_3^--N is not attracted to negatively charged particles of clay and organic matter. Rather, it is free to leach as water moves through the soil profile. **Figure 11-3** illustrates relative movement of NO_3^--N through different kinds of soils.

Recently, loss of N to the atmosphere as nitrous oxide (N_2O) has also been a concern because it is a greenhouse gas with 310 times the warming potential of carbon dioxide (CO_2) and is believed to be involved in global warming. Nitrous oxide is generated in soils by denitrification and nitrification.

All N sources (commercial, legumes, crop residues, soil organic mater, and animal manures) are readily converted to the NO_3^--N form in soils (see Chapter 3). Thus, all are subject to leaching into groundwater and loss as N_2O, unless utilized by a growing crop or retained in the ammonium-N (NH_4^+-N) form through proper management practices.

Sources of N may influence its environmental impact. For example, organic N sources often leave higher NO_3^--N levels in the soil because, based on current technology, their N release is more difficult to manage than is commercial fertilizer N. Among fertilizer N types, the source most effective in supplying plants

Soil Fertility Manual

often contributes less to soil N losses.

As described in Chapter 3, N undergoes numerous soil transformations depending on several factors, including moisture, temperature, soil pH, soil aeration, etc. The overall result is no net gain or loss of N in nature. The total process is known as the "nitrogen cycle" (see **Figure 3-1**).

Phosphorus

Phosphorus has been associated with environmental effects primarily through the euthrophication of lakes, bays, and non-flowing water bodies. Euthrophication is the response of a water body to over-enrichment by nutrients. That enrichment may be natural or man-made. Euthrophication symptoms are algal blooms, heavy growths of aquatic plants, algal mats, and hypoxia. Algal blooms in reservoirs can cause taste and odor problems in drinking water.

The fact that P is extremely immobile in soils was discussed in Chapter 4. It is held (adsorbed) very strongly by surfaces of iron (Fe), aluminum (Al), and manganese (Mn) oxides and hydroxides in acid soils. It is precipitated by calcium (Ca) ions to produce Ca phosphates of various types.

The nonpoint agricultural P additions to water bodies are mainly associated with erosion. Phosphorus movement is associated with erosion because:

- P has a very low solubility;

- P moves very little in most soils;

- Very low P concentrations are found in most drainage waters. Where erosion and sediment loss are stopped, P losses are minimized.

Phosphorus, along with N, is sometimes of concern with regard to its potential negative effects on surface water. However, contrary to the perception of some groups outside of agriculture, soils are not indiscriminately being built to excessive levels of P. In fact, a recent PPI survey, involving about 2.5 million soil samples, showed that 47% of the samples analyzed in North America were medium or below in P. State and provincial data are shown in **Production Concept 4-1**.

Excessive use of manures or fertilizer P can lead to soil buildup and a heightened risk of P loss to surface waters. Historically, P has been considered immobile in soils, but leaching can occur when the adsorption capacity of the soil is exceeded. There is evidence to suggest that the application of animal manures can result in greater downward movement of P through the soil than do mineral fertilizer sources.

Use of excessive amounts of animal manures is often associated with P buildup in mineral soils. Furthermore, soils which are used intensively for crops that require more P input than they remove—tobacco and potatoes, for example—may reach excessively high levels of P. As a result, P bound to soil sediments can pose a significant threat to nearby surface waters when erosion is a risk. The bulk of soil surface P loss is in suspended particles carried by runoff. Runoff also carries some dissolved P, while drainage water—particularly from tile drainage—can carry small amounts of both.

There are many interacting factors affecting the potential for P transport to surrounding landscapes. Often, soil test levels used alone do not provide adequate information to indicate the risk of P loss. Losses from erosion and runoff depend on several soil and landscape properties, such as: slope, soil cover, distance to surface water source, soil infiltration properties, soil test P, fertilizer P, fertilizer source (organic or mineral), fertilizer rate, and fertilizer placement.

To manage P applications and control P losses, the U.S. Department of Agriculture–Natural Resources Conservation Service (USDA-NRCS) has established a P Index (PI). The PI includes considerations that determine P source (s) and trans-

Table 11-1. An example of the P index developed by the U.S. Department of Agriculture.

Site characteristics (weighting factor)	Loss rating (value)				
	None (0)	Low (1)	Medium (2)	High (4)	Very High (8)
Transport Factors					
Soil erosion (1.5)	N/A	<5 t/A	5 to 10 t/A	10 to 15 t/A	>15 t/A
Irrigation (1.5)	N/A	Infrequent irrigation on well-drained soils	Moderate irrigation on soils with slopes <5%	Frequent irrigation on soils with slopes of 2 to 5%	Frequent irrigation on soils with slopes >5%
Soil runoff class (0.5)	N/A	Very low or low	Medium	Optimum	Excessive
Distance from watercourse (1.0)	>1,000 ft	1,000 to 500 ft	500 to 200 ft	200 to 30 ft	<30 ft
Source Factors					
Soil test P (1.0)	N/A	Low	Medium	Optimum	Excessive
P fertilizer application rate, lb P/A (0.75)	None applied	<15 lb P/A (35 lb P_2O_5/A)	16 to 40 lb P/A (36 to 92 lb P_2O_5/A)	41 to 65 lb P/A (93 to 149 lb P_2O_5/A)	>65 lb P/A (150 lb P_2O_5/A)
P fertilizer application method (0.5)	None applied	Placed with planter deeper then 2 in.	Incorporated immediately before crop	Incorporated >3 months before crop or surface applied >3 months before crop	Surface applied to pasture or applied <3 months before crop
Organic P source application rate, lb P/A (1.0)	None applied	< 15	16 to 40	41 to 65	>65
Organic P source application method (0.5)	None applied	Injected deeper than 2 in.	Incorporated immediately before crop	Incorporated >3 months before crop or surface applied >3 months before crop	Surface applied to pasture or applied <3 months before crop

Source: Sharpley, A.N., T. Daniel, T. Sims, J. Lemunyon, R. Stevens, and R. Parry. 1999. Agricultural phosphorus and eutrophication. U.S. Department of Agriculture, Agricultural Research Service, ARS-149, 42 p.

Soil Fertility Manual

portation processes that are unique to a given field or area. It allows farmers and their advisers to estimate loss potential for a specific site. It also offers management strategies to control P losses and to manage P fertilization to control soil test levels.

An example of a P index developed by the U.S. Department of Agriculture, Agriculture Research Service is shown in **Table 11-1**. Variations of this general index have been developed by several states. The index uses specific field features and management practices to obtain an overall rating for each field. Transport and source factors affecting P loss are weighted according to the assumption that certain characteristics have a relatively greater effect on potential P loss than others. Users must establish a range of P loss potential values for different geographic areas. Selecting the loss rating value for each site characteristic and multiplying it by the appropriate weighting factor assesses field vulnerability to P loss in surface runoff for that site characteristic. The P index for the field is the sum of the weighted values for all site characteristics. The site's general vulnerability to P loss can be categorized according to the P index value (**Table 11-2**).

Table 11-2. Generalized interpretation of the P index.

P index	General vulnerability to P loss
< 8	Low potential for P loss. If current farming practices are maintained, there is a low probability of adverse impacts on surface waters.
8 to 14	Medium potential for P loss. The chance for adverse impacts on surface waters exists, and some remediation should be taken to minimize the probability of P loss.
15 to 32	High potential for P loss and adverse impacts on surface waters. Soil and water conservation measures and P management plans are needed to minimize the probability of P loss.
> 32	Very high potential for P loss and adverse impacts on surface waters. All necessary soil and water conservation measures and a P management plan must be implemented to minimize the P loss.

Source: Sharpley, A.N., T. Daniel, T. Sims, J. Lemunyon, R. Stevens, and R. Parry. 1999. Agricultural phosphorus and eutrophication. U.S. Department of Agriculture–Agricultural Research Service, ARS-149, 42 p.

Where there is a risk of P loss that would impact on water quality, the PI is another tool that helps farmers fine tune their management systems to realize the greatest return on their investment while minimizing impact on the environment. Controllable management factors that increase P and other nutrient use efficiency include the following:

- nutrient management planning;
- timing of nutrient applications to meet crop needs and to avoid runoff-producing rainfall events;
- variable rate fertilizer application;
- conservation tillage
- grassland waterways and vegetative and riparian buffer zones
- incorporation of subsurface nutrient application.

Cadmium (Cd) is a heavy metal of some concern because of the levels found naturally in phosphate rock, the principle ore from which P fertilizers are made. Even though the long-term use of P fertilizers can result in a gradual buildup of Cd in the soil, it would take hundreds, perhaps thousands, of years at normal application rates to reach critical levels. The Association of American Plant Food Control Officials (AAPFCO) has proposed a tentative guide for metals in P and micronutrient fertilizers, shown in **Table 11-3**.

Table 11-3. Association of American Plant Food Control Officials (AAPFCO) Tentative Guide for Implementation under Section 12(a) of the Uniform State Fertilizer Bill. [1]

Metal	ppm per 1% P_2O_5	ppm per 1% micronutrients [2]	
Arsenic	13	112	
Cadmium	10	83	
Cobalt	3,100	23,000[3]	
Lead	61	463	
Mercury	1	6	
Molybdenum	42	300	[3]
Nickel	250	1,900	
Selenium	26	180[3]	
Zinc	420	2,900[3]	

URL=http://www.aapfco.org/#Start; ppm=parts per million

To use the table:

Multiply the percent guaranteed P_2O_5 or sum the guaranteed percentages of micronutrients (e.g. Zn) in each product by the value in the appropriate column in the table to obtain the maximum allowable concentration (ppm) of these metals. The minimum value for P_2O_5 utilized as a multiplier shall be 6.0. The minimum value for micronutrients uitilized as a multiplier shall be 1. If a product contains both P_2O_5 and micronutrients, multiply the guaranteed percent P_2O_5 by the value in the appropriate column and multiply the sum of the guaranteed percentages of the micronutrients by the value in the appropriate column. Utilize the sum of the two resulting values as the maximum allowable concentration.

Biosolids, and all compost products[4], shall be adulterated when they exceed levels permitted by the U.S. EPA Code of Federal regulations, 40 CFR Part 503. Dried biosolids and manure, as well as manipulated manure products, either separately or in combination, shall also be deemed adulterated when they exceed the levels permitted by the U.S. EPA Code of federal regulations, 40 CFR Part 503. Hazardous waste derived fertilizers (as defined by EPA) shall be adulterated when they exceed the levels of metals permitted by the U.S. EPA Code of Federal Regulations, 40 CFR, Parts 261, 266 and 268.

Notes:

[1] These guidelines are not intended to be used to evaluate horticultural growing media claiming nutrients, but may be applied to the sources of the nutrients added to the growing media.

[2] Micronutrients (also called minor or trace elements) are essential for both plant growth and development and are added to certain fertilizers to improve crop production and/or quality. These micronutrients are iron, manganese, zinc, copper, molybdenum, and boron. Nickel is also considered an essential micronutrient by some sources. In addition, cobalt and selenium can be considered micronutrients.

[3] Only applies when not guaranteed.

[4] Includes all compost products, separately or in combination with biosloids, manure, or manipulated manure, even those registered as fertilizers (making nutrient claims).

Potassium, Magnesium, and Sulfur

Potassium in water has no detrimental health or environmental effects. It is essential for human and animal health. Normal human dietary intake is from 2,000 to 6,000 milligrams of K per day...far above the content of water supplies. **Table 11-4** shows the effect K had on increasing corn silage yields and reducing the amount of carryover N.

Neither Mg nor S is considered to be of environmental concern from agricultural sources. Both are essential plant nutrients and must often be supplied by fertilization, based on soil testing and plant analysis. As with other essential nutrients, their deficiencies can decrease N and P use efficiency.

Table 11-4. Balance of N and K improves corn silage yields and N use efficiency.

N rate, lb/A	Without K			With K		
	Yield, tons/A	Soil N lb/A		Yield, tons/A	Soil N, lb/A	
		NO_3^-	NH_4^+		NO_3^-	NH_4^+
0	9.7	26	10.4	10.5	20	7.0
80	10.8	46	11.4	12.4	28	10.4
160	11.2	76	11.6	15.0	60	8.6

Quebec

Micronutrients

Micronutrients make significant contributions to food production and, thus, to human health. Applications of micronutrients, based on soil test or plant analysis, have a positive environmental impact through their effects on improved crop yields and more efficient use of other nutrients. Importance of micronutrients is increasing as crop yields increase and as sustained agricultural production requires that they be replaced in the soil.

There is some concern about the heavy metal contents of certain micronutrients manufactured from industrial by-products. It should be pointed out, however, that more than 97% of the mineral fertilizers manufactured in North America are made from natural sources such as atmospheric gases or mineral deposits rather than industrial by-products. At typical application rates, heavy metal loadings from biosolids—animal manures and sewage sludges—are generally higher than from mineral fertilizers. Further, contents of heavy metals (and micronutrients) in biosolids are highly variable. See **Table 11-3** for AAPFCO guidelines on heavy metals.

There is often confusion over chloride (Cl), one of the essential micronutrients. It has been confused with chlorine, which is a poisonous gas and never found free in nature. Chloride occurs in nature as sodium chloride (NaCl), potassium chloride (KCl), and salts of other metals. Chloride has not been associated with environmental or health problems. Potassium chloride—muriate of potash—is an important K fertilizer. It contains about 47% Cl. Sodium chloride—common table salt—is over 60% Cl.

Two distinct objectives must be considered to assure that adequate amounts of nutrients are used in agriculture for maintenance of profitable production levels while minimizing any potential negative effects on the environment.

Objective One: Balance all production inputs at optimum levels.

Objective Two: Utilize site-specific soil and water conservation techniques to provide optimum soil retention and minimum losses to ground-water.

Best management practices (BMPs) are necessary to attain both objectives. Best management practices involve both conservation and agronomic practices. Incorporating BMP technology into a cropping system plan is the key to both economical and environmental success. They are practices which have been proven in research and tested through farmer implementation to give optimum production potential, input efficiency, and environmental protection.

Best management practices help farmers achieve crop production yield levels where costs and NO_3^--N leaching per unit of production are near the minimum.

Where fertilizer N is applied at rates that do not exceed the economic optimum, its direct contribution to NO_3^- leaching is small for most field crops with extensive root systems, such as corn, wheat, and most other cereals. This may not be true for some high value horticultural crops on sandy soils.

Best management practice crop production, which includes adequate fertilization for optimum yield potential, will increase crop residues and the potential for NO_3^--N leaching from mineralization of N in the residues. But more residues also mean that soil organic matter levels are improved, which is a positive soil fertility and environmental factor. Management practices after

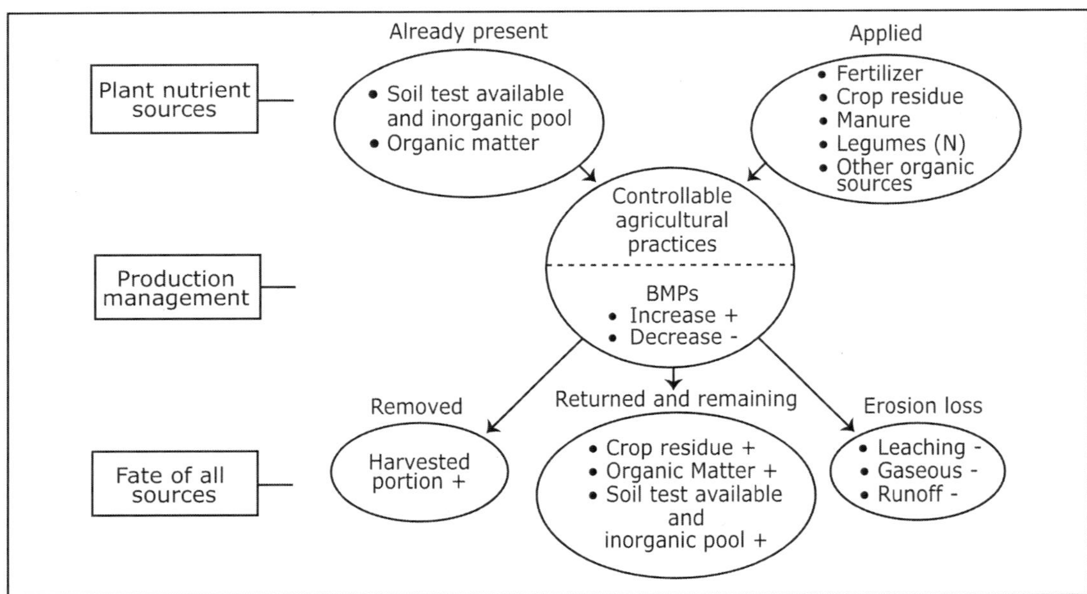

Figure 11.4. The relationship of nutrient sources, agricultural practices, and the fate of plant nutrients in a cropping system (adapted from Soil Science Society of America Special Publication 18).

Soil Fertility Manual

harvest to minimize NO_3^- –N leaching, such as the use of cover crops, are part of the BMP package.

Figure 11-4 provides a conceptual view of sources, agricultural practices, and fate of plant nutrients in a cropping system. Best management practices play a vital role in helping to increase nutrient use efficiency by the crop, to increase nutrient recycling in crop residues, and to build soil organic matter levels. At the same time, BMPs reduce nutrients lost via erosion, leaching, volatilization, denitrification, or runoff.

The environmental impact and agronomic response resulting from any one crop production input are determined as much, or more, by the management level of all the other controllable inputs in a production system as by the input itself.

An example of how this basic concept works for management of nutrients for corn production and environmental safety is presented in **Table 11-5**. Summaries from corn production research in several states show the impact various growth factors have on increasing yields and N efficiency. As these BMPs are put together to achieve an integrated crop management system, the efficiency of applied N and other nutrients improves

Table 11-5. The effects of several production inputs on corn yield and N efficiency from various locations.

Production factor	Yield, bu/A	Nitrogen efficiency, bu/lb N	State
Rotation:			
continuous	105	0.88	North Carolina
rotation	120	0.96	
Irrigation:			
without	127	0.51	New Jersey
with	214	0.86	
Planting date:			
late May	132	0.66	Indiana
early May	163	0.86	
Hybrid:			
bottom 5	149	0.60	
top 5	250	0.83	Florida
Population:			
low (12,000)	155	0.52	Florida
high (36,000)	231	0.96	
Compaction:			
compacted	123	0.62	Indiana
not compacted	167	0.84	
pH x P:			
low pH, low P	90	0.60	Wisconsin
best pH, high P	138	0.92	
P Placement (row):			
no P_2O_5	143	0.64	
35 lb P_2O_5	159	0.71	
70 lb P_2O_5	165	0.73	Wisconsin
K x Planting date:			
no K x late	125	0.52	Illinois
no K x early	142	0.59	
K x late	156	0.65	
K x early	170	0.71	

while the possibility of any detrimental effects on water quality decreases.

A BMP fertilizer recommendation allows crop yields to be expressed at the economic optimal level, which for most crops is also the point of greatest environmental protection.

An important BMP for N is identifying the correct rate required to maximize yield. When N rate exceeds crop N need, residual NO_3^-–N may leach if water is sufficient to move below the root zone. However, when N rate is matched to yield potential, there is little danger of NO_3^--N leaching deep into the soil profile. In a long-term Oklahoma study, various N rates were applied to wheat on plots located in four areas of the state. Soil NO_3^--N accumulation at depths below the top 12 in. of plots receiving recommended N rates was not significantly different from NO_3^--N levels in those plots that did not receive N. In other words, optimum yields were produced while environmental risks were minimized. **Figure 11-5** shows the results from one of the experimental locations.

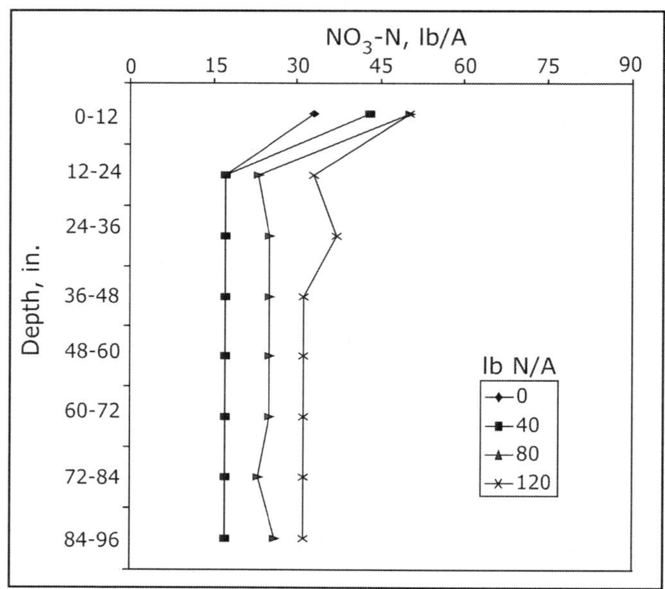

Figure 11-5. Soil NO_3^--N in lb/A per profile increment as a function of N applied, following 18 years of annual applications in continuous wheat. Recommended N rate was 40 lb/A. (Adapted from Westerman et al. Agron J. 86: 94-99.)

Remember that nutrient recommendations obtained from a laboratory represent an estimate based on previous field experiments and thus should be adjusted for the specific field and management situation.

1) Set yield goals: Determine realistic goals for each crop and each field.

Nutrient requirements increase with yield. **Table 11-6** shows the nutrients taken up by wheat at three yield levels. It is a BMP to be sure that adequate, but not excessive, nutrients are readily available to the growing crop from seeding to maturity.

Table 11-6. Nutrient requirements for wheat increase at higher yields.

Yield, bu/A	Nutrient uptake, lb/A				
	N	P_2O_5	K_2O	Mg	S
60	125	41	138	18	15
80	166	54	184	24	20
110	229	75	253	33	28

2) Use soil testing and plant analysis.

These are the best tools available to determine the levels and availability of soil nutrients and the amounts of nutrients which should be applied to achieve the yield goal. Monitor nutrient needs with frequent soil tests, especially for P, K, S, Mg, micronutrients, and pH. Plant analysis helps confirm diagnosis of nutrient needs and can identify needs during the growing season (see **Chapter 9**).

In addition, more reliable soil N tests are being developed. Tissue analysis for N is a good tool to use to help determine the amount of N to apply during the growing season when more accurate rates and more efficient utilization of applied N can be expected.

Table 11-7 shows that N efficiency was almost doubled by P applications on a medium P testing soil in Kansas. This is significant in the major wheat states where from one-half to three-fourths of the soils test medium or less in P. Similar observations can be cited for cotton, corn, rice, and other crops which require N fertilizer and where a response to P, K, S, Mg, micronutrients, or aglime is expected.

Table 11-7. Adequate P increases wheat yields and improves N use efficiency.

Rate, lb/A		Yield, bu/A	Nitrogen efficiency, bu/lb N	N balance sheet soil (-); unused (+) lb/A
N	P_2O_5			
- - - lb/A - - -				
75	0	35	0.46	+ 9
75	20	51	0.68	- 21
75	30	56	0.75	- 30
75	40	61	0.81	- 40
75	50	64	0.84	- 45

Kansas

3) Follow a conservation plan.

Best management practices for soil and water are site-specific. They include: conservation tillage, terracing, contour stripcropping, grass waterways, contour and grass headlands, crop rotation and cover crops, and water control basins and diversions.

A good soil and water conservation plan for each farm can be the single most important factor in decreasing erosion and the potential loss of soil, water, and nutrients, especially P attached to sediment and organic particles.

Some form of conservation tillage can be practiced in almost all types of farming systems. **Table 11-8** shows the positive effects of conservation tillage in reducing water runoff and the associated sediment and P losses. Decreased runoff means greater water infiltration and more available water for the growing crop. Total P losses are greatly reduced through lower sediment losses.

Table 11-8. Conservation tillage reduces runoff volume, sediment, and P losses.

Tillage treatment	Runoff volume, gal/A	Sediment, lb/A	Total P lost, lb/A
Conventional	30,510	138	0.16
No-till	4,999	37	0.01

Maryland

4) Adopt BMPs of all controllable inputs to achieve higher yields.

Well-fertilized soils, along with other good management practices, lead to higher yields (see **Table 11-4**). Higher yields, with their associated increase in crop residues, have a tremendous positive effect in reducing water runoff and water and wind erosion losses. **Table 11-9** shows that as crop residue was increased from a quarter-ton to one ton per acre, soil loss decreased 10-fold.

Table 11-9. Increased residue reduces soil loss.

Surface residue, t/A	Soil loss, t/A
0.25	3.0
1.00	0.3

Indiana

Optimum soil fertility levels have many benefits.

- A more rapid crop canopy closure results from better plant nutrition. It reduces the erosive energy of raindrops, improves moisture use efficiency, and reduces weed pressure. Phosphorus available at early growth stages is especially important for early canopy development.

- Vigorous growth of plants both above and below ground helps to hold soil in place, improves water infiltration and water use efficiency (WUE), and increases yields. **Table 11-10** provides one example of the effects of good fertility management on corn yields and WUE.

Table 11-10. Effect of balanced fertility and corn yields on WUE.

Fertility level	Yield, bu/A	WUE, bu/in.[1]
Low	76	3.1
Medium	149	6.0
High	240	9.6

[1]25 inches of water

Florida

5) Application timing plans.

Nitrogen efficiency and yield potential are increased when applications are made close to the time of greatest crop uptake. Consider split N applications at growth stages which correspond to the N demands of the crop. Where irrigation is used, N application through the irrigation system is an efficient application method. On some coarse-textured (sandy) soils, split applications of K, S, and some micronutrients may also be a BMP because yield potential is increased and environmental protection is enhanced through more efficient utilization of other inputs.

The NRCS recommends the following precautions to minimize loss if fall N applications must be applied:

- Use a fertilizer that contains only NH_4^+-N.

- Incorporate or band the fertilizer into the soil.

- Wait until soil temperatures at the surface 4 in. depth remain below 50°F.

- Use a nitrification inhibitor.

6) Inhibitors.

- **Nitrification inhibitors** retard the conversion of NH_4^+-N to NO_3^--N. When used with the appropriate rates of commercial N fertilizer or animal manure, they can increase N uptake by the crop. The inhibitors hold N in the soil in the stable NH_4^+ form, which is not subject to leaching or other losses. Ammonium stays in the crop rooting zone so it is available for crop uptake, even under wet soil conditions that move NO_3^- deeper into the soil and out of the reach of crop roots. Research shows that corn, wheat, cotton, grain sorghum, and many other crops use NH_4^+-N readily and tend to take up more total N when NH_4^+-N is available along with NO_3^-.

  Leaching studies with nitrification inhibitors and stabilizers have shown reductions in NO_3^--N leaching from 8 to 27%. In addition to the environmental benefit, nitrification inhibitors increase yield potential and the efficient use of N applications, both economic benefits (**Table 11-11**).

- **Urease inhibitors** (e.g. NBPT, Agrotain®) are effective risk-management tools in slowing the breakdown of urea into its components: ammonia (NH_3) and NH_4^+. They improve long-term N use efficiency by reducing the amount of N lost to the atmosphere, so more is

Table 11-11. Examples of nitrification inhibitor (nitrapyrin) effects on grain yield and N use efficiency.

N Source	Minnesota (corn) None	Minnesota (corn) Inhib.	Illinois (corn) None	Illinois (corn) Inhib.	Nebraska (corn) None	Nebraska (corn) Inhib.	Kansas (sorghum) None	Kansas (sorghum) Inhib.
	- - - bu/A - - -		- - - bu/A - - -		- - - bu/A - - -		- - - bu/A - - -	
Ammonia	161	172	184	194	–	–	–	–
Urea	147	165	–	–	–	–	–	–
N solution	145	163	–	–	135	143	100	110

recovered by the crop. Also, by slowing the conversion to NH_3 and NH_4^+, urease inhibitors reduce potential seedling toxicity. Both NH_3 and NH_4^+ are toxic to seedlings in high concentrations.

7) Fertilizer placement plans.

Higher crop yields are not only a matter of fertilizer rates and methods of application, but fertilizer placement as well. Proper fertilizer placement improves nutrient availability to plants, which means higher yields and greater nutrient efficiency, particularly in conservation tillage systems.

Starters, banding, strip or dribble, deep injection, and knifing are just some of the terms used for various methods of fertilizer placement. The increased use of conservation tillage systems has increased the need to consider the most efficient methods of applying fertilizer nutrients. For example, the injection or banding of N in conservation tillage systems helps to prevent tie-up of nutrients with the surface residue, minimizes the volatilization losses of ammonia N, and improves positional availability. New fertilizer equipment technology makes it possible to place nutrients correctly with increased accuracy and with a minimum of delay.

Early growth response to band or row applications of fertilizer, especially P, is commonly observed. Responses are most frequent in cold climates or when crops are planted in cooler soils. These responses often occur soils high in P. An example of a row P fertilizer response is presented in **Table 11-12**.

Table 11-12. Effect of soil test P and row P on corn yield.

Broadcast P_2O_5 spring applied, lb/A	Soil test P, lb/A Spring	Soil test P, lb/A Fall	Row P_2O_5, lb/A 0	Row P_2O_5, lb/A 20	Row P_2O_5, lb/A 40	Row P_2O_5, lb/A 60
			- - - - Corn yield, bu/A - - - -			
0	60	70	103	137	134	139
137	65	62	119	134	144	132
687	112	122	122	142	149	141

Wisconsin

Overall, fertilizer placement has advantages because of:

- Possible delayed fertilizer P and K reactions in the soil due to diminished fertilizer-soil contact.

- Deeper placement of nutrients into soil where moisture is less limiting to uptake.

- Less tie-up of P with soil components due to higher concentrations of NH_4^+-N in the soil P retention zone.

- Forced plant uptake of NH_4^+-N, causing a more acid condition at the root surfaces favoring P absorption.

- A better fit in conservation tillage systems because of the difficulty of incorporating nutrients adequately and because of residue immobilization of nutrients.

Summary

In the final analysis, the use of adequate amounts of plant nutrients for optimum crop yields and profitability is a key to environmental protection. Best management practices developed through research, modified and adopted for specific site conditions, are a key to both efficient use of nutrients and protection of our soil and water resources. SFM

Other Resources for Soil Fertility Information

Aglime Facts. 1996. Potash & Phosphate Institute/Foundation for Agronomic Research. Norcross, GA.

Better Crops with Plant Food. Quarterly publication of the International Plant Nutrition Institute. Available on the website at:>**www.ipni.net**<.

Foundation for Agronomic Research website: >**www.farmresearch.com**<

International Plant Nutrition Institute website: >**www.ipni.net**<

Micronutrients in Agriculture, 2nd edition. Mortvedt, J.J., F.R. Cox, L.M. Shuman, and R.M. Welch, eds. 1991. Soil Science Society of America. Madison, WI. 1991.

Nitrogen in Agricultural Soils. 1982. American Society of Agronomy. Madison, WI.

Nutrient Deficiencies and Toxicities in Crop Plants. 1993. Bennett, W.F., ed. The American Phytopathological Society. St. Paul, MN.

Plant Nutrient Use in North American Agriculture. 2001. PPI/PPIC/FAR Special Publication 2001-1. Potash & Phosphate Institute.

Potassium in Agriculture. 1985. Munson, R.D., ed. American Society of Agronomy, Madison, WI.

Preparing for the International Certified Crop Adviser Exam. 2007. Gilmour, J.T. International Plant Nutrition Institute.

Soil Fertility and Fertilizers: An Introduction to Nutrient Management. Seventh Edition. Prentice Hall. 1999. Havlin, J.L., J.D. Beaton, S.L. Tisdale, and W.L. Nelson. Prentice Hall.

Soil Test Levels in North America. PPI/PPIC/FAR Special Publication 2005-1. 2006. Fixen, PE. Potash & Phosphate Institute.

Site-Specific Management Guidelines. PPI/PPIC/FAR. Website: >**www.ppi-far.org/ssmg**<.

The Role of Phosphorus in Agriculture. Khasawneh, F.E., E.C. Sample, and E.J. Kamprath, eds. 1980. American Society of Agronomy.

USDA/ARS website: >**www.ars.usda.gov**<

USDA/NRCS website: >**www.nrcs.usda.gov**<

Western Fertilizer Handbook. Ninth Edition. 2002. California Plant Health Association. Interstate Publishers, Inc.

Recognizing Nutrient Deficiency Symptoms in Crops

The classic signs of nutrient deficiency (or toxicity) symptoms are not usually found in well-managed crop fields today. With good management and modern production practices, conditions resulting in deficiency problems are avoidable. Even so, knowledge of those symptoms and underlying causes can be important in diagnosis of problems. In some areas, cutback in application of nutrients in recent years is now showing up as deficiency symptoms due to depleted soils.

In general, if deficiency symptoms do appear on crops during the growing season, significant yield loss has already occurred. Crops are more likely to suffer from "hidden hunger" or conditions which tend to limit yields or quality without apparent symptoms. More than one nutrient deficiency may occur at the same time. Symptoms may occur in a limited area within a field or sometimes over the entire field. Occurrence can be related to such properties as poor drainage, sandy conditions, infertile zones, or areas treated differently in the past. If a symptom is found in a single plant, consider disease, injury, or a genetic variation. Also, earlier symptoms are often more useful to note than those that develop as the plant matures.

Keep in mind that some nutrients are relatively "immobile" in plants, while others are more "mobile." Typically, the symptoms caused by deficiencies of an "immobile" nutrient will occur on the upper or younger leaves. The older leaves will remain green and free of symptoms because these immobile nutrients do not move or translocate from them.

The deficiency symptoms shown on these pages are only a sampling. The descriptions are general and may not fit all circumstances.

Nitrogen deficiency in young corn causes the entire plant to be pale and yellowish green, with spindly stalks. Later, V-shaped yellowing may appear on the tips of leaves. Nitrogen is a mobile nutrient in the plant. Yellowing begins at the leaf tip, along the midrib on lower, older leaves, and progresses up the plant if the deficiency persists.

Phosphorus-deficient corn plants are usually smaller and grow more slowly than plants with adequate P. They may be dark green with reddish purple tips and leaf margins. The deficiency is usually identified on young plants. Phosphorus is readily mobilized and translocated in the plant.

Potassium deficiency on corn may appear as yellowing and necrosis of the leaf margins, beginning on the lower leaves. If the deficiency persists, the leaf symptoms will progress up the plant. Potassium-deficient corn plants age too fast, cells die, and tissues deteriorate, inviting stalk rot.

Sulfur deficiency on small corn plants may appear as a general yellowing of the foliage, similar to N deficiency. Yellowing of the younger leaves is more pronounced than with N deficiency because S is not easily translocated in the plant. Problems are more likely on acid, sandy soils, or low organic matter soils.

In soybeans, P deficiency symptoms may not be well defined. Soybeans require large amounts of P, especially at pod set. Phosphorus-deficient plants are spindly, with small leaflets and retarded growth. Leaves may appear dark green or bluish-green. Leaf analysis for P content is the best way to diagnose deficiency.

Potassium deficiency in soybeans is well defined. Soybeans require large amounts of K. It is important for all aspects of plant growth and influences the plant's nutritional balance. It is also involved in the uptake of calcium and magnesium. Potassium deficiency symptoms appear first on older leaves. In early stages of growth, an irregular yellow mottling appears around leaflet margins.

Sulfur-deficient soybean plants are pale green, with the youngest leaves often appearing more yellow. Stems are thin, hard, and elongated, with small, yellow-green leaves at the top of the plant. Sulfur deficiency may lead to reduction of protein synthesis. Availability of soil S depends on the rate at which it is released from organic matter, influenced by plant residues, soil moisture, temperature, and soil pH.

Manganese deficiency in soybeans may appear as pale green, interveinal chlorosis of newer leaves.

Nitrogen deficiency in wheat and other small grains may first appear as yellowing and then as stunted growth. Chlorosis usually begins on older tissues such as lower leaves. Cell growth and division as well as protein synthesis may be slowed. Wheat and other small grains and grasses are generally sensitive to insufficient N and are responsive to supplemental N applications.

Phosphorus-deficient wheat plants sometimes maintain their green color and may be darker green than plants with sufficient P. However, they are slow-growing and slow to mature. Tillering is often reduced or lacking completely. Leaf tips die back when shortages are severe, and foliage of some varieties may show shades of purple or red. Older leaves and older tissues are the first to show P deficiency symptoms.

Symptoms of S deficiency in wheat and other small grains are similar to those of N deficiency. Sulfur deficiency is more common in mineral soils that are well drained, coarse-textured, and low in organic matter. Sulfur deficiency in wheat typically appears first on younger tissues, but eventually causes the entire plant to take on a pale green appearance.

Chloride deficiency in wheat (sometimes referred to as physiological leaf spot) can be a problem in some regions, such as the Great Plains. Lesions typically appear at flag leaf emergence and are more severe on the tip half of the leaf. Symptoms are first apparent on older (lower) leaves and progress to successively younger leaves over a two or three week period.

Symptoms of P deficiency in alfalfa may appear as stunted plant growth. Older leaves may be small and exhibit dark green to purple discoloration. Deficient leaflet appears at left above, with a normal leaflet at right.

Potassium deficiency in alfalfa may appear as two different symptoms. Typically, spots develop on margins of more mature leaves and may coalesce to produce entire yellow margins. A second symptom is a white chlorosis at the margins and toward the leaf tips associated with high sodium contents. The two symptoms do not appear on the same plant.

Magnesium deficiency in corn can be confused with Fe deficiency. Symptoms include interveinal yellowing or white discoloration beginning with lower leaves, as Mg is translocated from older to newer leaves.

Nitrogen deficiency symptoms in grain sorghum may include chlorosis on older, lower leaves first as N is translocated to newer, developing leaves. The yellowing proceeds from the leaf tip down the midrib in a V-shaped pattern. Eventually, necrosis or "firing" of the entire lower leaves occurs as the plant matures.

Iron deficiency in grain sorghum results in severely stunted plants with interveinal chlorosis of entire leaves occurring with newer leaves first. Leaves turn white under severe Fe stress.

Zinc deficiency in corn may appear as bleached or pale yellow discoloration in the area between the leaf edge and midrib.

Copper deficiency in corn shows symptoms of pale yellow to white discoloration of newer leaves, progressing to necrosis of leaf tips and edges.

Copper deficiency in wheat can be mistaken for leaf diseases. Symptoms include chlorosis and necrosis of newer leaves, with twisting of leaf tips to give a hooking appearance.

Boron deficiency symptoms in white clover, shown here, include orange-red and pale green leaves. Symptoms may also include red leaf margins, leather leaf texture, and poor seed head development.

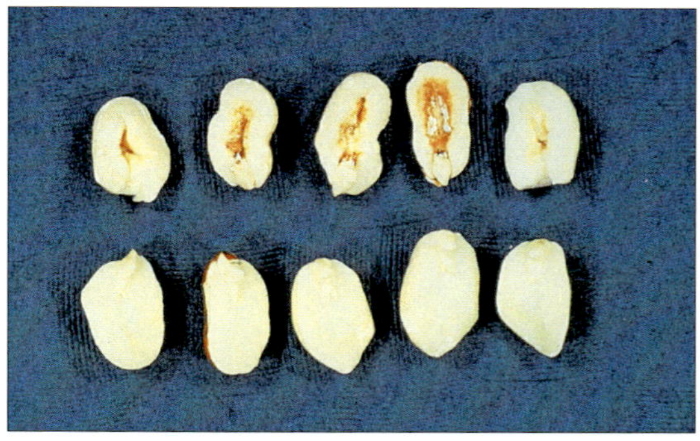

Boron deficiency in peanuts causes the condition called hollow heart, shown in the top row of this photo.

Phosphorus deficiency in potatoes results in plants appearing stunted and darker green than normal. As the severity of the deficiency increases, the leaves roll upward, which exposes the gray-green lower surface, giving the field a normal green color again.

Potato plants suffering from K deficiency may show symptoms of marginal leaf scorching, followed by necrosis and browning. Young, fully expanded leaves become crinkly and develop a glossy sheen with a slightly black pigmentation.

Nitrogen deficiency symptoms on cotton early in the season include yellowish green leaf color, first appearing on older leaves. Younger leaves may be reduced in size. Plant height is also reduced, few vegetative branches develop, fruiting branches are short, and bolls may be shed soon after flowering. When N deficiency occurs later in the season on plants with a moderate load of maturing bolls, foliar symptoms appear as reddening in the middle of the canopy.

Foliar symptoms of K deficiency in cotton that occur before peak bloom may include interveinal light green to gold mottling first on older leaves, with yellowing and necrosis developing at leaf margins under severe deficiency. Late-season K deficiency results in foliar symptoms that differ from early season deficiency.

In canola, S deficiency appears as interveinal chlorosis apparent on much of the plant. Newly developed leaves are small, and growth is stunted. Severe deficiency results in failure to set seed.

Calcium is an important nutrient for tomatoes. Blossom end-rot is one consequence of deficiency. The problem can occur even in calcareous soil due to inadequate translocation of Ca within the plant for rapid-growing and high-yielding varieties.

Magnesium deficiency in broccoli affects older leaves first as chlorosis between veins.

Potassium deficiency in apples is characterized by a scorching of leaf margins. Margins may first appear light green and later necrotic or tattered. In some cases, K deficiency may reduce the color or acidity of fruit.

Zinc deficiency on citrus (orange) affects younger leaves first. Small, narrow leaves have yellow mottling between the veins.

Manganese deficiency in celery produces chlorotic leaves with dark green veins at the top of the plant. Symptoms first appear on the newer (upper) leaves. Deficiency is more likely if the soil pH is high and on soils higher in organic matter during cool spring months when soils are waterlogged.

Phosphorus deficiency symptoms on winegrapes may vary on different varieties. Reddish spots may appear randomly across leaves. Entire leaves may become chlorotic as deficiency progresses. Symptoms in some varieties may include broad reddish blotches across green leaves or symptoms concentrated along leaf margins.

Symptoms of K deficiency in grapes typically appear in early summer on the middle portion of the shoots. Leaves become chlorotic beginning at the margins and progressing inward between the veins. Leaves tend to cup downward. In white wine varieties, leaves become yellow or yellow-bronze. Red varieties show stronger bronzing and reddening.

Locations of Nutrient Deficiency Symptoms

Nutrient	Position on plant	Chlorosis?	Leaf margin necrosis?	Color and leaf shape
N	All leaves	Yes	No	Yellowing of leaves and leaf veins
P	Older leaves	No	No	Purplish patches
K	Older leaves	Yes	Yes	Yellow patches
Mg	Older leaves	Yes	No	Yellow patches
Ca	Young leaves	Yes	No	Deformed leaves
S	Young leaves	Yes	No	Yellow leaves
Mn Fe	Young leaves	Yes	No	Interveinal chlorosis
B, Zn, Cu, Ca, Mo	Young leaves	—	—	Deformed leaves

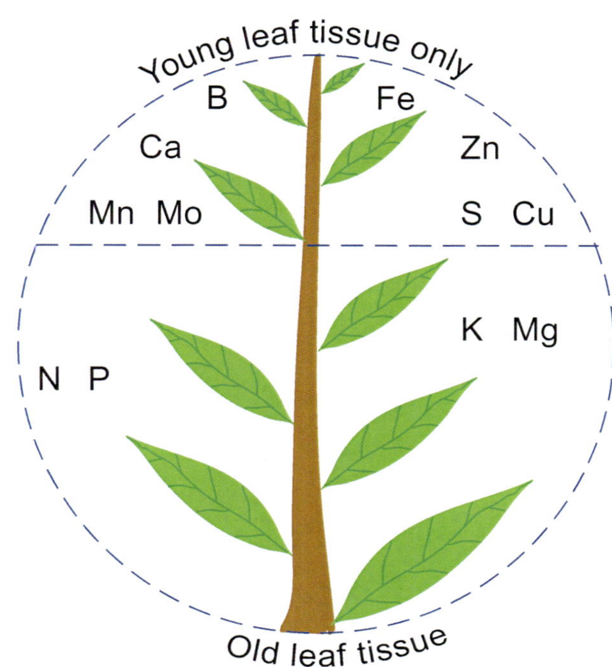

Nitrogen (N) deficiency

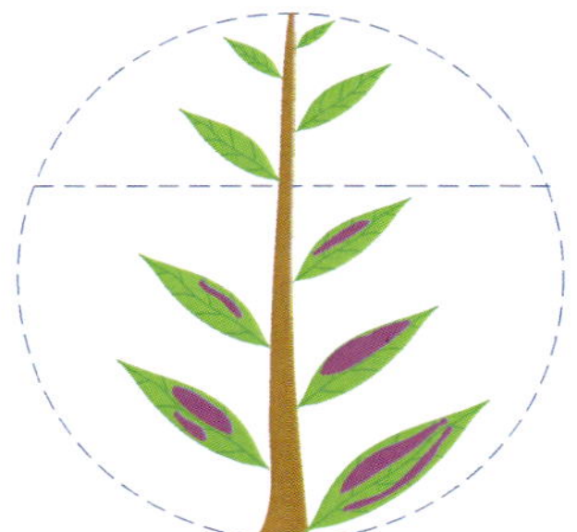

Phosphorus (P) deficiency

Potassium (K) deficiency

Manganese (Mn) and iron (Fe) deficiency

Sulfur (S) deficiency

Magnesium (Mg) deficiency

Micronutrient (Ca, B, Mo, and Zn) deficiencies

For more information, check the website at
>**www.ipni.net**<.

Useful Conversions, Tables, Charts, and Reference Lists

Contents

B-1.   Conversion Factors for U.S. System and Metric Units.

To convert column 1 into column 2, multiply by	Column 1	Column 2	To convert column 2 into column 1, multiply by
Length			
0.621	kilometer, km	mile, mi	1.609
3.281	meter, m	foot, ft.	0.3048
1.094	meter, m	yard, yd	0.914
0.394	centimeter, cm	inch, in.	2.54
Area			
0.386	kilometer^2, km^2	mile^2, mi^2	2.590
247.1	kilometer^2, km^2	acre	0.00405
2.471	hectare, ha	acre	0.405
Volume			
35.32	cubic meters, m^3	cubic feet, ft^3	0.02832
0.00973	meter^3, m^3	acre-in.	102.8
0.0284	liter, l	bushel, bu	35.24
1.057	liter, l	quart (liquid), qt	0.946
Mass			
1.102	tonne (metric ton)	ton (short)	0.9072
2.205	quintal, q	hundredweight, cwt (short)	0.454
2.205	kilogram, kg	pound, lb	0.454
0.035	gram, g	ounce (avdp), oz	28.35
Pressure			
14.50	bar	lb/in.^{2}, psi	0.06895
0.9869	bar	atmosphere, atm	1.013
0.9678	kg (weight)/cm^2	atmosphere, atm	1.013
14.22	kg (weight)/cm^2	lb/in.^{2}, psi	0.07031
14.70	atmosphere, atm	lb/in.^{2}, psi	0.06805
Yield or Rate			
0.446	tonne (metric ton)/ha	ton (short)/acre	2.240
0.891	kg/ha	lb/acre	1.12
0.891	quintal/hectare	hundredweight/acre	1.12
1.15	hectoliter/hectare, hl/ha	bu/acre	0.870
Temperature			
(1.8 x C) + 32	Celsius, C	Fahrenheit, F	0.56 (F-32)
	-17.8°C	0°F	
	0°C	32°F Freezing point of water	
	20°C	68°F	
	37°C	98.6°F Human body temperature	
	100°C	212°F Boiling point of water	

Metric Prefix Definitions

mega	1,000,000	deca	10	centi	0.01
kilo	1,000	basic metric unit	1	milli	0.001
hecto	100	deci	0.1	micro	0.000001

Conversion examples for U.S. system and metric units:

To convert 40 lb/A of P_2O_5 to its metric equivalent — 40 lb/A x 1.12 = 44.8 kg/ha

To convert 70 tonnes/ha of sugarbeets to its U.S. system equivalent — 70 t/ha x 0.446 = 31.2 tons/A

Other conversion factors for U.S. crop yields to metrics:

Corn — bu/A x 0.0629 = tonnes/ha

Soybeans — bu/A x 0.0676 = tonnes/ha

Wheat — bu/A x 0.0676 = tonnes/ha

Grain Sorghum — bu/A x 0.0629 = tonnes/ha

B-2. Conversions and Equivalent Values[1].

Fluid Measures

1 fluid ounce	=	29.5 milliliters
1 fluid ounce	=	2 tablespoons or 6 teaspoons
8 fluid ounces	=	16 tablespoons or 1 cup
16 fluid ounces	=	1 pint or 2 cups
32 fluid ounces	=	1 quart or 2 pints
128 fluid ounces	=	1 gallon or 4 quarts

1 gallon of water weighs 8.34 pounds
1 gallon of fuel oil weighs 7.2 pounds

Avoirdupois Weight

16 drams	=	1 ounce or 437.5 grains
16 ounces	=	1 pound
2,000 pounds	=	1 short ton
2,240 pounds	=	1 long ton (1 metric ton = 2,204.6 lb)

Troy Weight

24 grains (gr)	=	1 penny weight (pwt or dwt)
20 pennyweight	=	1 ounce
12 ounces	=	1 pound (lb)

Apothecaries Weight

20 grains	=	1 scuple (sc)
3 scruples	=	1 dram (dr)
8 drams	=	1 ounce (oz)
12 ounces	=	1 pound (lb)

Household Measure

60 drops	=	1 teaspoon or $\frac{1}{6}$ ounce
2 teaspoons	=	1 dessert spoon
3 teaspoons	=	1 tablespoon or $\frac{1}{2}$ ounce
16 tablespoons	=	1 cup
2 cups	=	1 pint or 16 fluid ounces

millimhos per centimeter (mmhos/cm) = deciSiemens per meter (dS/m)
millimhos per centimeter (mmhos/cm) x 1,000 = micromhos per centimeter

1 ppm = 1 mg/kg = 1 mg/L=0.0001% = l lb/million lb
1% (w/v) = 10 g/L = 1.33 oz/gal = 8.34 lb/100 gal (U.S.)
                1.60 oz/gal = 10.02 lb/100 gal (Imperial)
1% (v/v = 1.28 fl oz/gal (U.S.) = 1.60 fl oz (Imperial) = 1 gal/100 gal
1 cu ft. water = 7.48 gal. (U.S.) = 62.4 lb = 28.32 L
1 mmhos/cm = 1,000 μmhos/cm = 1 dS/cm
1 meq/L = 1 mmol$_c$/L = 1 mol/m^3
1 meq/100 g = 1 mmol$_c$/100g = 1 cmol$_c$/kg

[1]Imperial measure used in Canada
1 fluid ounce (fl. oz.) = 28.413 milliliters
1 pint (pt.) = 20 fl. oz. = 0.5683 liters
1 gallon (gal) = 8 pt. = 4.5461 liters

Length

Inch = 1/12 or 0.083 foot = 2.54 centimeters = 25.4 millimeters
Foot = 12 inches = 0.3048 meters = 30.48 centimeters
Yard = 36 inches = 3 feet = 0.9144 meters
Furlong = 220 yards
Mile = 1,760 yards = 5,280 feet = 1.61 kilometers = 8 furlongs = 80 chains

Area

Square inch = 0.00694 square foot = 6.45 square centimeters
Square foot = 144 square inches = 929.03 square centimeters
Square yard = 9 square feet = 0.836 square meters
Acre = 4,840 square yards = 43,560 square feet = 160 square rods =
    4,047 square meters = 0.405 hectare
Hectare = 10,000 square meters = 2.47 acres
Square mile = 640 acres = 2.59 square kilometers = 1 section
Section = 1 square mile = 640 acres = 2.59 square kilometers

Liquid Measures

Teaspoon = 0.1667 fluid ounce = 60 drops = 4.93 milliliters
Tablespoon = 3 teaspoons = 0.5 fluid ounce = 14.8 milliliters
Fluid ounce = 2 tablespoons = 29.58 milliliters
Cup = 8 fluid ounces = 16 tablespoons = 236.6 milliliters
Pint = 2 cups = 16 fluid ounces = 473.2 milliliters
Quart = 4 cups = 2 pints = 32 fluid ounces = 0.946 liters
Liter = 2.113 pints = 1,000 milliliters = 1.057 quarts
*Gallon = 4 quarts = 8 pints = 128 fluid ounces = 3.785 liters
Cubic foot of water = 7.5 gallons = 62.4 pounds = 28.3 liters
Acre inch of water = 27,154 gallons = 3,630 cubic feet
Imperial gallon = 4.8 quarts = 9.6 pints = 154 fluid ounces = 4.55 liters

Dry Measures

Teaspoon (level) = 0.35 cubic inch = 5.74 cubic centimeters
Tablespoon (level) = 1.05 cubic inch = 3 level teaspoons = 17.21 cubic centimeters
Cup = 16 level tablespoons = 16.8 cubic inches = 275.3 cubic centimeters
Pint = 2 cups = 32 level tablespoons = 33.6 cubic inches = 550.6 cubic centimeters
Quart = 2 pints = 64 tablespoons = 67.2 cubic inches = 1.101 liters
Peck = 8 quarts = 16 pints = 538 cubic inches = 8.8 liters
Bushel = 4 pecks = 2,150 cubic inches = 32 quarts = 35 liters

Volumes

Cubic inch = 0.000579 cubic foot = 16.4 cubic centimeters
Cubic foot = 1,728 cubic inches = 0.0370 cubic yard = 0.0283 cubic meter
Cubic yard = 27 cubic feet = 0.765 cubic meters

Weights

Gram = 15.43 grains = 1,000 milligrams
Ounce = 28.35 grams = 437.5 grains
Pound = 16 ounces = 7,000 grains = 454 grams
Kilogram = 1,000 grams = 2.205 pounds
Ton (short) = 2,000 pounds = 0.907 metric tons (tonnes)
Ton (long) = 2,240 pounds = 1.016 metric tons (tonnes)

*To find the number of gallons in a square or oblong tank, multiply the number of cubic feet it contains by 7.4805.
To find the number of gallons in a circular tank or well, square the diameter in feet, multiply by the depth in feet, and then multiply by 5.875.

B-4. Plant Population per Acre.

Spacing	Population	Spacing	Population
1 by 2 ft.	21,780	6 by 6 ft.	1,210
1 by 4 ft.	10,890	8 by 8 ft.	680
1½ by 3 ft.	9,680	12 by 12 ft.	302
2 by 4 ft.	5,445	16 by 16 ft.	170
3 by 5 ft.	2,904	20 by 20 ft.	109
4 by 4 ft.	2,722	30 by 30 ft.	48

Determining Seeding Rate per Square Foot

Row width, in.	\- \- \- Desired seeds per square foot \- \- \-								
	8	12	16	20	24	28	32	36	40
	- - - - - - - - - - - Seeds per linear foot of row - - - - - - - - - - -								
6	4	6	8	10	12	14	16	18	20
7	5	7	9	12	14	16	19	21	23
8	5	8	11	13	16	19	21	24	27
9	6	9	12	15	18	21	24	27	30
10	7	10	13	17	20	23	27	30	33
11	7	11	15	18	22	26	29	33	37
12	8	12	16	20	24	28	32	36	40
13	9	13	17	22	26	30	35	39	43
14	9	14	19	23	28	33	37	42	47
15	10	15	20	25	30	35	40	45	50
16	11	16	21	27	32	37	43	48	53
17	11	17	23	28	34	40	45	51	57
18	12	18	24	30	36	42	48	54	60

How to Check Corn Stands Quickly

To estimate thousands of stalks per acre:
1) Measure or step off length of row equal to 1/1,000 acre, as shown in the table below.
2) Count number of stalks and multiply by 1,000. Check several sections of row and take average.

Row width, in.	Row length for 1/1,000 acre
20	26 ft., 1 in.
22	23 ft., 9 in.
30	17 ft., 4 in.
36	14 ft., 6 in.
40	13 ft., 1 in.

An example: For 30 in. rows, measure 17 ft., 4 in. or step off about six paces. Count back. If you have 26 stalks, that means a population of about 26,000 per acre.

B-5. Application Rates.

The following data are useful in calculating rates of application:

1 acre-foot of soil = 4,000,000 lb (approximate)

1 acre-foot of water = 325,851 gal = 1,233.5 cu m

1 t per acre = 20.8 g per sq. ft.

1 t per acre = 1 lb per 21.78 sq. ft.

1 t per acre = 25.12 quintals per hectare

1 t per acre 6-in. depth – 1 g per 1,000 g of soil

1 g per sq. ft. = 95.9 lb per acre

1 lb per acre= 0.0104 g per sq. ft.

1 lb per acre = 1.121 kg per hectare

100 lb per acre = 0.2296 lb per 100 sq. ft.

g per sq. ft. x 95.9 = lb per acre

kg per 48.0 sq. ft. = t per acre

lb per sq. ft. x 21.78 = t per acre

lb per sq. ft. x 43,560 = lb per acre

lb per 1,000 sq. ft. x 0.4878 = kg per 100 sq m

lb per cu yd x 0.0005932 = g per cc

lb per cu yd x 0.05932 = kg per cu m

100 sq. ft. = 1/435.6 or 0.002296 acre

t per acre-foot = 0.00136 x parts per million

cu ft. per second = 448.83 gal per minute

parts per million = 17.1 x grains per gal

parts per million x 0.00136 = t per acre-foot

Commodity	Unit	Approx. net weight, lb
Alfalfa seed	Bushel	60
Almond	lb (in shell)	1
Apples	Bushel basket or carton	40
Avocados	Lug	12-15
	Flat or carton, 2 layer	26
Barley	Bushel	48
Beans:		
Lima, dry	Bushel	56
Other, dry	Bushel	60
Snap	Bushel	28-32
Bluegrass seed	Bushel	14-30
Cabbage	Open mesh back, sack, wirebound crate	50
Carrots, without tops	Sacks, 48 1-lb and 24 1-lb	48
	Sacks	50
Celery	Caron or crate	60
Corn:		
Ear, husked	Bushel	70
Shelled	Bushel	56
Sweet	Carton	42
Cotton	Bale, gross	500
	Bale, net	480
Cottonseed	Bushel	32
Flaxseed	Bushel	56
Grapefruit:		
Florida and Texas	½-box mesh bag	40
California and Arizona	Carton	34
Grapes, western	Lug	28
Lettuce	Carton	50
Millet	Bushel	48-60
Oats	Bushel	32
Peanuts, unshelled		
Virginia	Bushel	17
Spanish	Bushel	25
Pears, California	Carton	36
Peas, green		
Unshelled	Bushel	28-30
Dry	Bushel	60
Peppers, green	Carton	28
Pistachio	lb (in shell)	1
Plums	½-bushel carton	28
Prunes	½-bushel carton	30
Potatoes	Sack	100
Rice	Bushel	45
Sorghum grain	Bushel	56
Soybeans	Bushel	60
Strawberries	12, 1 pint	12
Sweet potatoes	Carton	40
Tomatoes	Carton	25
	2 layer flat	20
Watermelon	Carton	85
Wheat	Bushel	60

Source: Weights, Measures, and Conversion Factors for Agricultural Commodities and Their Products, Economic Research Service, Agricultural Handbook No. 697.

B-7.   Atomic Weights and Common Valence Values.

Name	Symbol	Atomic weight	Common valence
Aluminum	Al	26.98	3
Boron	B	10.81	3
Calcium	Ca	40.08	2
Carbon	C	12.01	-4, 4
Chlorine	Cl	35.45	-1
Cobalt	Co	58.93	2
Copper	Cu	63.55	2
Fluorine	F	19.00	-1
Hydrogen	H	1.01	-1, 1
Iodine	I	126.90	-1
Iron	Fe	55.85	2, 3
Magnesium	Mg	24.31	2
Manganese	Mn	54.94	2, 4
Molybdenum	Mo	95.94	2, 6
Nickel	Ni	58.69	2
Nitrogen	N	14.01	-3 to 5
Oxygen	O	16.00	-2
Phosphorus	P	30.97	5
Potassium	K	39.10	1
Sodium	Na	23.00	1
Sulfur	S	32.07	-2, 4, 6
Zinc	Zn	65.38	2

B-8.   Chemical Symbols, Equivalent Weights, and Common Names of Ions, Salts, and Chemical Amendments.

Chemical Symbol or Formula	Gram Equivalent Weight	Common Name
Ca^{2+}	20.04	Calcium ion
Mg^{2+}	12.15	Magnesium ion
Na^+	23.00	Sodium ion
K^+	39.10	Potassium ion
Cl^-	35.46	Chloride ion
NO_3^-	62.01	Nitrate ion
NH_4^+	17.03	Ammonium ion
SO_4^{2-}	48.03	Sulfate ion
CO_3^{2-}	30.00	Carbonate ion
HCO_3^-	61.02	Bicarbonate ion
$CaCl_2$	55.50	Calcium chloride
$CaSO_4$	68.07	Calcium sulfate
$CaSO_4 \cdot 2H_2O$	86.09	Gypsum
$CaCO_3$	50.04	Calcium carbonate
$MgCl_2$	47.62	Magnesium chloride
$MgSO_4$	60.19	Magnesium sulfate
$MgCO_3$	42.16	Magnesium carbonate
$NaCl$	58.46	Sodium chloride
Na_2SO_4	71.03	Sodium Sulfate
Na_2CO_3	53.00	Sodium carbonate
$NaHCO_3$	84.02	Sodium bicarbonate
KCl	74.56	Potassium chloride
K_2SO_4	87.13	Potassium sulfate
K_2CO_3	69.10	Potassium carbonate
$KHCO_3$	100.12	Potassium bicarbonate
S	16.03	Sulfur
SO_2	32.03	Sulfur dioxide
H_2SO_4	49.04	Sulfuric acid
$Al_2(SO_4)_3 \cdot 18H_2O$	111.08	Aluminum sulfate
$FeSO_4 \cdot 7H_2O$	139.02	Iron sulfate (ferrous)

B-9. Salt Index (Relative Effect of Fertilizer Materials on the Soil Solution)[1].

Material	Salt Index	Partial Salt Index per Unit of Plant Nutrient
Anhydrous ammonia	47.1	0.572
Ammonium nitrate	104.7	2.990
Ammonium nitrate-lime	61.1	2.982
Ammonium phosphate (11-48-01)	26.9	2.442
Ammonium polysulfide (20-0-0-40S)	43.6	2.180
Ammonium sulfate	69.0	3.253
Ammonium thiosulfate (12-0-0-26S)	84.4	7.040
Calcium carbonate (limestone)	4.7	0.083
Calcium cyanamide	31.0	1.476
Calcium nitrate	52.5	4.409
Calcium sulfate (gypsum)	8.1	0.247
Diammonium phosphate	29.9	1.614 (N)
		0.637 (P_2O_5)
Dolomite (calcium and magnesium carbonates)	0.8	0.042
Kainit, 13.5%	105.9	8.475
Kainit, 17.5%	109.4	6.253
Manure salts, 20%	112.7	5.636
Manure salts, 30%	91.9	3.067
Monoammonium phosphate (11-52-0)	34.2	2.453 (N)
		0.485 (P_2O_5)
Monocalcium phosphate	15.4	0.274
Nitrate of soda (sodium nitrate)	100.0	6.060 (N)
Nitrogen solution, 37%	77.8	2.104
Nitrogen solution, 40%	70.4	1.724
Potassium chloride, 50%	109.4	2.189
Potassium chloride, 60%	116.3	1.936
Potassium chloride, 63%	114.3	1.812
Potassium nitrate	73.6	5.336 (N)
		1.580 (K_2O)
Potassium polysulfide (0-0-22-23S)	37.7	1.720
Potassium sulfate	46.1	0.853 (K_2O)
Potassium thiosulfate (0-0-25-17S)	64.0	2.560
Sodium chloride	153.8	2.899 (Na)
Sulfate of potash-magnesia	43.2	1.971 (K_2O)
Superphosphate, 16%	7.8	0.487
Superphosphate, 20%	7.8	0.390
Superphosphate, 45%	10.1	0.224
Superphosphate, 48%	10.1	0.210
Urea	75.4	1.618
Urea-ammonium nitrate solution (32-0-0)	95.0	2.304

[1]After L.F. Rader, Jr., et al., Soil Sci., 55:21-218, 1943.

B-10. Part I.   Nutrient Content of Some Organic Materials.

Material	N	P$_2$O$_5$	K$_2$O	Ca	Mg	S	Cl
			Percentage by weight				
Apple pomace	0.2	—	0.2	—	—	—	—
Blood (dried)	12 to 15	3	—	0.3	—	—	0.6
Bone meal (raw)	3.5	22	—	22	0.6	0.2	0.2
Bone meal (steamed)	2	28	0.2	23	0.3	0.1	—
Brewers grains (wet)	0.9	0.5	—	—	—	—	—
Compost (garden)	varies with components and amendments						
Cotton waste from factory	1.3	0.4	0.4	—	—	—	—
Cottonseed hull ash	0	—	27	—	—	—	—
Cottonseed meal	6 to 7	2.5	1.5	0.4	0.9	0.2	—
Cowpea forage	0.4	0.1	0.4				
Eggs	2.2	0.4	0.2	—	—	—	—
Egg shells	1.2	0.4	0.2	—	—	—	—
Feathers	15.3	—	—	—	—	—	—
Fermentation sludges	3.5	0.5	0.1	7.3	0.1	—	—
Fish scrap (acidulated)	5.7	3	—	6.1	0.3	0.2	0.5
Fish scrap (dried)	9.5	6	—	6.1	0.3	0.2	1.5
Fly ash:							
coal	0.3	—	0.1	0.48	—	—	—
wood	0.1	0.6	10	9.8	0.66	—	—
Greensand	—	1 to 2	5	—	—	—	—
Grape skins (ash)	—	3.6	31	—	—	—	—
Hair	12 to 16	—	—	—	—	—	—
Hay							
legume	3	1	2.4	1.2	0.2	0.3	—
grass	1.5	0.5	1.9	0.8	0.2	0.2	—
Milk	0.5	0.3	0.2	—	—	—	—
Peanut hull meal	1.2	0.5	0.8	—	—	—	—
Peanut meal	7.2	1.5	1.2	0.4	0.3	0.6	0.1
Peat/muck	2.7	—	—	0.7	0.3	1	0.1
Pine needles	0.5	0.1	—	—	—	—	—
Poultry processing:							
DAF sludge	8	1.8	0.3	—	—	—	—
Potato tubers	0.4	0.2	0.5	—	—	—	—
Potato, leaves & stalks	—	0.6	0.2	0.4	—	—	—
Sawdust	0.2	—	0.2	—	—	—	—
Seaweed (dried)	0.7	0.8	5	—	—	—	—
Sewage sludge (municipal)	2.6	3.7	0.2	1.3	0.2	—	—
Shrimp waste	2.9	10	—	—	—	—	—
Soot from chimney flues	—	0.5 to 11	—	1	0.4	—	—
Soybean meal	7	1.2	1.5	0.4	0.3	0.2	—
Spanish moss	0.6	0.1	0.6	—	—	—	—
Spent brewery yeast	—	7	0.4	0.3	0.04	0.03	—
Sweet potato skins boiled (ash)	—	3.29	13.9	—	—	—	—
Sweet potatoes	0.2	0.1	0.5	—	—	—	—
Tankage	7	1.5	3 to 10	—	—	—	—
Textile sludges	2.8	2.1	0.2	0.5	0.2	—	—
Wood ashes	0	2	6	20	1	—	—
Wood processing wastes	—	0.4	0.2	0.1	1.1	0.2	—
Tobacco stalks	3.7	0.6	4.5	—	—	—	—
Tobacco stems	2.5	0.9	7	—	—	—	—
Tomatoes, fruit	0.2	0.1	0.4	—	—	—	—
Tomato leaves	0.4	0.1	0.4	—	—	—	—

Note: Approximate values are given. Have materials analyzed for nutrient content before using.
Source: North Carolina State University website.

B-10. Part II. Nutrient Content of Some Manures.

Type	TKN	P_2O_5	K_2O	Ca	Mg	S
			- Wet basis -			
Dairy						
Fresh (lb/t)	10	5	8	4	2	1
Paved surface scraped (lb/t)	10	6	9	5	2	2
Liquid manure (lb/1,000 gal)[1]	23	14	21	10	5	3
Lagoon liquid (lb/acre-in.)[2]	137	77	195	69	35	25
Anaerobic lagoon sludge (lb/acre-in.)[2]	15	22	8	12	4	4
Beef						
Fresh (lb/t)	12	7	9	5	2	2
Paved surface scraped (lb/t)	14	9	13	5	3	2
Unpaved feedlot (lb/t)	26	16	20	14	6	5
Lagoon liquid (lb/acre-in.)[2]	83	77	129	24	19	—
Lagoon sludge (lb/1,000 gal)[1]	38	51	15	36	5	—
Broiler						
Fresh (lb/t)	26	17	11	10	4	2
House litter (lb/t)	72	78	46	41	8	15
Stockpiled litter (lb/t)	36	80	34	54	8	12
Duck						
Fresh (lb/t)	28	23	17	—	—	—
House litter (lb/t)	19	17	14	22	3	3
Stockpiled litter (lb/t)	24	42	22	27	4	6
Goat						
Fresh (lb/t)	22	12	18	—	—	—
Horse						
Fresh (lb/t)	12	6	12	11	2	2
Layers						
Fresh (lb/t)	26	22	11	41	4	4
Undercage paved (lb/t)	28	31	20	43	6	7
Deep pit (lb/t)	38	56	30	86	6	9
Liquid (lb/1,000 gal)[1]	62	59	37	35	7	8
Lagoon liquid (lb/acre-in.)[2]	179	46	266	25	7	52
Lagoon sludge (lb/1,000 gal)[1]	26	92	13	71	7	12
Rabbit						
Fresh (lb/t)	24	23	13	19	4	2
Sheep						
Fresh (lb/t)	21	10	20	14	4	3
Unpaved (lb/t)	14	11	19	24	7	6
Swine						
Fresh (lb/t)	12	9	9	8	2	2
Surface scraped (lb/t)	13	12	9	12	2	2
Liquid manure (lb/1,000 gal)[1]	31	22	17	9	3	5
Lagoon liquid (lb/acre-in.)[2]	136	53	133	25	8	10
Lagoon sludge (lb/1,000 gal)[1]	22	49	7	16	4	8
Turkey						
Fresh (lb/t)	27	25	12	27	2	—
House litter (lb/t)	52	64	37	35	6	9
Stockpiled litter (lb/t)	36	72	33	42	7	10

Notes: Approximate nutrient contents are given. Have materials analyzed for nutrient content before using. North Carolina mean waste analysis 1981 to 1990 supplied by J.C. Barker, NCSU Department of Biological and Agricultural Engineering.
[1]Pounds per thousand gallons of manure liquid (slurry). Corrected from pounds per thousand pounds (lb/1,000 lb) in original source. [2]Pounds per acre-inch. Estimated total lagoon liquid includes total liquid manure plus average annual lagoon surface rainfall surplus; does not account for seepage.
Source: North Carolina State University website.

Soil Fertility Manual

B-11. Nutrient Sources – Fertilizers.

Name	Formula	Raw material	Source of raw materials/Process
Urea	$CO(NH_2)_2$	NH_3 and CO_2	Atmosphere/'Once Through', 'Partial' and 'Total Recycle' processes, based on Haber-Bosch reaction
Ammonium sulfate (AS)	$(NH_4)_2SO_4$	Chemical by-products	NH_3, $CaSO_4$, CO_2, Calcium ammonium nitrate/Saturation of H_2SO_4 with NH_3, Leuna process
Calcium ammonium nitrate (CAN)	$NH_4NO_3 + CaCO_3$	Chemical by-product	HNO_3, NH_3, $CaCO_3$/$Ca(NO_3)_2$, Combustion of NH_3 + $CaCO_3$/ODDA process
Triple superphosphate	$Ca(H_2PO_4)_2$	Terrestrial	P rock, H_3PO_4/Treatment of pulverized P rocks with phosphoric acid
Diammonium phosphate	$(NH_4)_2HPO_4$	Chemical by-products	NH_3, H_3PO_4/Wet-process reaction (acid slurry + NH_3 at 115°C)
Phosphate rock (rock phosphate)	$Ca_{10}(PO_4)_6F_2$	Terrestrial	Phosphate ore/Purrification and grinding
(Double) Superphosphate	$Ca(H_2PO_4)_2+CaSO_4$	Terrestrial	P rock, H_2SO_4, (H_3PO_4)/Treatment of pulverized P rocks with sulfuric and phosphoric acid
Fused magnesium phosphate (FMP)	$Ca_{10}(PO_4)_6F_2 + Mg_3H_4Si_2O_9$	Terrestrial	Phosphate ore and serpentine/ Fusion of pulverized raw materials in furnaces.
Muriate of potash, (MOP)	KCl	Marine-lacustrine	KCl, $MgCl_2$, $NaCl$, etc./Thermal dissolution, flotation, crystallization
Sulfate of potash	K_2SO_4	Marine-lacustrine	Rock salt/Thermal dissolution, flotation, crystallization
Langbeinite	$K_2SO_4 \cdot MgSO_4$	Marine-lacustrine	Rock salt/Segregation, granulation
Kieserite	$MgSO_4 \cdot H_2O$	Marine-lacustrine	Rock salt $K_2Mg(SO_4)_2$/Segregation, electrostatic benefications, granulation
Dolomite	$CaMg(CO_3)_2$	Marine sediments	Calcite/dolomite/Segregation, grinding
Borate (sodium borate)	$Na_2B_4O_{13} \cdot 4H_2O$	Lacustrine	Ulexite, borax/Segregation of precipitated lacustrine salts
Copper sulfate	$CuSO_4 \cdot 5H_2O$	Marine-lacustrine	$CuFeS_2$, FeS_2, $CuFeS_4$/Reaction of Cu melt products + H_2O

B-12. Conversion Table for Nutrient Concentrations in Compounds.

To convert column 1 to column 2, multiply by	Column 1	Column 2	To convert column 2 to column 1, multiply by
4.43	N	NO_3	0.23
1.22	N	NH_3	0.82
4.72	N	$(NH_4)_2SO_4$	0.21
2.86	N	NH_4NO_3	0.35
1.20	K	K_2O	0.83
0.63	KCl	K_2O	1.58
0.71	CaO	Ca	1.40
0.56	$CaCO_3$	CaO	1.78
0.60	MgO	Mg	1.66
2.99	MgO	$MgSO_4$	0.33
3.43	MgO	$MgSO_4 \cdot H_2O$	0.29
2.09	MgO	$MgCO_3$	0.48
0.16	$MgSO_4 \cdot 7H_2O$	MgO	6.25
2.29	P	P_2O_5	0.44
0.46	$Ca_3(PO_4)_2$	P_3O_3	2.18
2.00	S	SO_2	0.50
2.50	S	SO_3	0.40
3.00	S	SO_4	0.33
0.13	$MgSO_4 \cdot 7H_2O$	S	7.68
0.25	$(NH_4)_2SO_4$	S	4.00
0.23	$MgSO_4 \cdot H_2O$	S	4.31

B-13. Chemical Symbols for Essential Plant Nutrients and Common Forms Absorbed by Crops.

Nutrient	Symbol	Common forms absorbed by crops[1]
Nitrogen	N	Nitrate (NO_3^-); ammonium (NH_4^+)
Phosphorus	P	Primary orthophosphate ($H_2PO_4^-$); secondary orthophosphate (HPO_4^{2-})
Potassium	K	K^+
Sulfur[2]	S	Sulfate (SO_4^{2-})
Calcium	Ca	Ca^{2+}
Magnesium	Mg	Mg^{2+}
Boron	B	Boric acid, H_3BO_3
Chloride	Cl	Cl^-
Copper	Cu	Cu^+; Cu^{2+}
Iron	Fe	Fe^{2+}
Manganese	Mn	Mn^{2+}
Molybdenum	Mo	Molybdate (MoO_4^{2-})
Nickel	Ni	Ni^{2+}
Zinc	Zn	Zn^{2+}

[1]Plants get their carbon (C), hydrogen (H), and oxygen (O) from air and water.
[2]Sometimes absorbed through the leaves as SO_2 gas.

The values presented in the following table are estimates of nutrient removal or the quantity of nutrients removed in the harvested portion of the crop. They should not be confused with nutrient uptake, which refers to the total nutrients absorbed by the growing crop. Tabular values are approximations based on the most recent information available to PPI. Actual nutrient removal may vary by 30% or more depending on the specific growing conditions of the crop such as soil fertility level, yield, soil moisture, crop vigor, and limiting nutrients (interactions) as well as the actual crop variety and fertilizer program. Changes to soil fertility may differ from the amount removed by the crop. In some instances, weathering of soil minerals and organic matter may compensate for part of the nutrient removal by crops. In other conditions, nutrients may be chemically fixed by the soil or lost by leaching, and the loss of nutrients will exceed crop removal.

B-14. Nutrients Removed in Harvested Portion of Crop.

Type of crop	Unit	N	P_2O_5	K_2O
Field Crops				
Barley (spring)	lb/bu	1.10	0.40	0.35
Canola	lb/bu	1.88	0.91	0.46
Corn (grain)	lb/bu	0.75	0.44	0.29
Corn (grain)	lb/cwt	1.34	0.79	0.52
Corn (silage, 67% water)	lb/t	8.30	3.60	8.30
Cotton (lint)	lb/bale	32.00	14.00	19.00
Flax	lb/bu	2.00	1.10	0.65
Lentils	lb/bu	2.00	0.62	1.10
Oats	lb/bu	0.80	0.25	0.20
Peanuts[1]	lb/t	70.00	11.00	17.00
Peas (field)	lb/bu	2.40	1.20	0.71
Rice	lb/cwt	1.27	0.67	0.35
Sorghum (grain)	lb/cwt	1.50	0.75	0.38
Soybeans[1]	lb/bu	4.00	0.80	1.40
Sugarbeets	lb/t	4.00	1.50	6.60
Sugarcane	lb/t	2.00	1.25	3.50
Sunflower	lb/cwt	2.80	1.10	0.60
Tobacco (flue)	lb/cwt	2.80	0.50	5.20
Tobacco (burley)	lb/cwt	4.30	0.43	4.70
Wheat: 10% protein[2]	lb/bu	1.10	0.50	0.35
12% protein[2]	lb/bu	1.30	0.50	0.35
14% protein[2]	lb/bu	1.50	0.50	0.35
Forage Crops (Dry matter basis)				
Alfalfa[1]	lb/t	56	15	60
Bermudagrass	lb/t	46	12	50
Bromegrass	lb/t	36	13	59
Clover[1]-grass	lb/t	50	15	60
Fescue	lb/t	38	18	52
Orchardgrass	lb/t	50	17	62
Sorghum-sudan	lb/t	40	15	58
Timothy	lb/t	38	14	62
Vetch[1]	lb/t	56	15	46
Vegetable Crops				
Broccoli	lb/cwt	0.44	0.17	0.42
Cabbage	lb/cwt	0.39	0.09	0.36
Lettuce	lb/cwt	0.24	0.08	0.50
Potatoes	lb/cwt	0.35	0.15	0.56
Squash	lb/cwt	0.42	0.10	0.60
Sweet potatoes	lb/cwt	0.52	0.23	1.00
Tomatoes	lb/t	2.50	0.92	5.70
Fruit and Nut Crops				
Almonds (in shell)	lb/t	130	50	170
Apples	lb/t	6.0	3.6	16.8
Cantaloupe	lb/t	7.3	2.3	13.0
Grapes (table)	lb/t	8.3	3.0	13.0
Oranges	lb/t	8.8	1.8	11.0
Peaches	lb/t	6.3	2.7	8.0

[1] Legumes obtain most of their N from the air.   [2] At same moisture content as yield measurement.
Source: Plant Nutrient Use in North American Agriculture. 2002. PPI/PPIC/FAR Technical Bulletin 2002-1.

Index

SOIL FERTILITY MANUAL

The *SOIL FERTILITY MANUAL,* first published in 1978, was revised and updated in 2003. It still features the basic principles of soil/plant relationships and fertilizer/aglime use. The Manual serves as a practical and modern resource on soil fertility, including the primary and secondary nutrients and micronutrients, soil sampling and testing, plant analysis, other diagnostic techniques, and more.

The publication now contains 11 chapters, plus a glossary and appendix sections containing tables of conversions and reference lists, and color photos of various nutrient deficiency symptoms in crops. The new version totals about 200 pages and has an index of topics.   The presentation style of the manual is practical and easy-to-grasp, making it adaptable to use by agronomists, farm advisers, students, farmers, and others...suitable for classroom use, meetings, and individual study. ISBN # 0-9629598-5-5

Quantity   Cost

Soil Fertility Manual purchase information

Item # 50-5200   **Cost: $25.00** (complete with cover, wire-o binding)   _______ $ __________

Item # 50-5100   **Cost: $20.00** (without cover, 3-hole punched)   _______ $ __________

CD-ROM with PowerPoint files for 11 chapters

Compact disk contains almost 400 images with concepts and illustrations related to chapter topics of the Soil Fertility Manual.

Item # 82-6500   **Cost: $50.00**   _______ $ __________

Subtotal **US$**__________

Shipping/handling–see other side **US$**__________

Total = **US$**__________

SHIPPING ADDRESS: (Please Print**)**

Name _______________________________________

Company ___________________________________

Address ____________________________________

City _____________ State/Prov. _______________

Country _____________ Zip/Postal Code __________

Phone_____________________Fax________________

E-mail _____________________________________

BILLING ADDRESS: (If different from above**)**

Name _______________________________________

Company ___________________________________

Address ____________________________________

City _____________ State/Prov. _______________

Country _____________ Zip/Postal Code __________

Phone_____________________Fax________________

E-mail _____________________________________

Discounts for quantity orders of a single item:			
50-99	10%	200-299	20%
100-199	15%	300 up	25%

METHOD OF PAYMENT:

❏ Check Enclosed (include shipping/handling–see other side) Payable to "International Plant Nutrition Institute"

Credit cards accepted:

❏ Discover   ❏ VISA

❏ MasterCard ❏ American Express

Enter card number

Signature _________________________________

Exp. Date________________

Send order to: Circulation Department, IPNI
3500 Parkway Lane, Suite 550
Norcross, GA 30092-2844
Phone: (770) 825-8082 or 825-8084

Or fax your order to: (770) 448-0439

E-mail: circulation@ipni.net   Reference # 08034

Date:__________________   Federal ID# 53 0026300

Check the web site at
www.ipni.net

IPNI
INTERNATIONAL
PLANT NUTRITION
INSTITUTE

Shipping / Handling Charges

For orders sent to IPNI prepaid, please include the shipping/handling fee with payment.

Follow these charts to figure shipping/handling. Over $200.00 please contact Norcross office.

ORDERS TO BE DELIVERED IN THE UNITED STATES CANADA AND MEXICO

Total value of materials ordered:	U.S.	CANADA	MEXICO
Up to $15.00	$6.00	$12.00	$12.00
$16.00 to $45.00	$10.00	$15.00	$19.00
$46.00 to $100.00	$14.00	$19.00	$27.00
$101.00 to $150.00	$18.00	$26.00	$32.00
$151.00 to $200.00	$21.00	$30.00	$38.00

ORDERS TO BE DELIVERED IN SOUTH & CENTRAL AMERICA

Total value of materials ordered:	Amount for shipping/handling:
Up to $15.00	$14.00
$16.00 to $50.00	$21.00
$51.00 to $100.00	$29.00
$101.00 to $150.00	$34.00
$151.00 to $200.00	$41.00

ORDERS TO BE DELIVERED IN AUSTRALIA PACIFIC RIM AFRICA

Total value of materials ordered:	Amount for shipping/handling:
Up to $15.00	$14.00
$16.00 to $50.00	$26.00
$51.00 to $100.00	$32.00
$101.00 to $150.00	$38.00
$151.00 to $200.00	$44.00

ORDERS TO BE DELIVERED IN EUROPE EASTERN EUROPE ASIA

Total value of materials ordered:	Amount for shipping/handling:
Up to $15.00	$14.00
$16.00 to $50.00	$20.00
$51.00 to $100.00	$26.00
$101.00 to $150.00	$34.00
$151.00 to $200.00	$40.00

IPNI
INTERNATIONAL
PLANT NUTRITION
INSTITUTE

Suite 550
3500 Parkway Lane
Peachtree Corners, Georgia 30092-2844
U.S.A.
Phone: (770) 447-0335
www.ipni.net